THE JAWS
OF DEATH

THE JAWS OF DEATH

SHARKS AS PREDATOR, MAN AS PREY

XAVIER MANIGUET

TRANSLATED BY DAVID A. CHRISTIE

Skyhorse Publishing

www.skyhorsepublishing.com

10 9 8 7 6 5 4 3 2 1

Library of Congress Cataloging-in-Publication Data

Maniguet, Xavier.
  [Dents de la mort. English]
  The jaws of death : sharks as predator, man as prey / Xavier
Maniguet.
      p. cm.
  Originally published: Dobbs Ferry, NY : Sheridan House, 1994.
  ISBN-13: 978-1-60239-021-8 (alk. paper)
  ISBN-10: 1-60239-021-5 (alk. paper)
  1. Shark attacks. 2. Sharks—Behavior. I. Title.

QL638.93.M3613 2007
597.3—dc22

                              2007018588

Printed in Canada

CONTENTS

CONTENTS

PREFACE

It is an honour and a pleasure to have been invited by my friend Xavier Maniguet to write the foreword for his book on sharks. A quick look through the list of contents will convince both researchers and amateurs with an interest in sharks that the book is a real mine of information. It will stimulate research and at the same time destroy numerous myths and many pseudo-scientific speculations.

Xavier Maniguet's visit to the Natal Shark Board in South Africa, in March 1989, was one of the great moments of the quarter-century I have spent in the bosom of this organisation. It is rare that one meets somebody as well qualified in so many disciplines. He is as well informed on all spheres of man's activity as he is on the habits of sharks. This being so, we exchanged long and fruitful discussions on the subject of these magnificent animals

For many, the word "shark" conjures up horrifying visions of formidable, man-eating machines. The media are, alas, largely responsible for this misguided attitude. For example, before the first *Jaws* film was released, shark attacks, whether fatal or of a minor nature, barely received a mention in the American press. Since then, the slightest incident immediately makes the front page of the newspapers. I really do not know of any other creature that has been so maligned.

My association (or love affair) with sharks goes back to 1st March 1966, when I was put in charge of the Natal Anti-Shark Measure Board. This board had been created by a special order in 1964, in response to a general revolt by the population following a series of seven attacks between December 1957 and summer 1958. The attacks resulted in five deaths and ruined the tourist industry. As a result, total war was declared against all sharks. The Natal Anti-Shark Measure Board was given official responsibility for implementing and monitoring measures for safeguarding bathers against attacks by sharks.

At Durban, the municipal authorities, in the wake of twenty-one attacks between 1943 and 1951, had, in 1952, wisely decided to follow the example of the Australians by installing nets off its main beaches. There was no further attack even though the number of bathers increased each year. In view of such successful results, the other big bathing resorts demanded that the same installations be provided for them. Thus when I took up my post, I inherited 87 nets off sixteen beaches to the north and south of Durban. In fact I inherited a system which evidently worked 100 per cent effectively, but without anybody being able to explain why! There was no other solution for me than to go out on the boats that hauled up the nets every morning in order to answer this question and many others. All the nets have to be set up and maintained from outboard boats, and these are obliged to cross the surf on both journeys, there and back. The danger is not insignificant, since this surf is known for being one of the most powerful on the planet.

At the start, the boats were rudimentary, far removed from the sophistication of our fleet of today, and it was with my lifebelt that I very often boarded at dawn to accompany the working parties. I very soon discovered that the sharks get caught in the nets at night, and very rarely during the daytime. The statistics show that 65% of catches are made on the landward side. The sharks have a diurnal periodicity of swimming, moving towards the coast in surface waters once night has fallen, and leaving again at deeper levels for the open sea at dawn. These habits account for our notice-boards on all protected beaches, advising people not to enter the water at night "when shark attack is likely".

As a zoologist, I have always regarded shark nets as an excellent means of study. The teams have to fill out cards which show the species, the sex and the length of each shark caught, as well as the side of the net where the shark was found, the temperature of the water, etc. For each shark captured, 70 pieces of information are recorded. Thanks to these highly qualified teams, the NSB has at its command a databank on sharks which is the largest in the world.

It has always saddened me that sharks have to be sacrificed on the altar of tourism. The nets in fact protect an industry of tens of millions of South African Rand against animals which have existed for millions of years and which are perfectly adapted to their environment, rightfully holding their position as greatest predators and lords of our oceans. It is man who is the intruder in the domain of these magnificent creatures.

I am proud that, thanks to vigorous education programmes, the attitude of the young, and the not so young, has advanced greatly. In 1985, the name of the organisation was, at my request, changed by parliamentary ruling to "The Natal Shark Board".

When I met Dr Perry W. Gilbert, doyen of American researchers, in 1969, at the Sarasota Mote Marine Laboratories, Florida, I remember him saying: "You know, we are very lucky to be working on sharks. They take us to magnificent places and bring us in contact with some fascinating people." Xavier Maniguet is one of these.

Beulah Davis, Director, Natal Shark Board, Durban

ACKNOWLEDGEMENTS

I thank all those who assisted me on diving trips all over the world, and helped me to get close to this magnificent animal the shark. In particular I thank my friends in the Republic of South Africa who have given me access to their studies and authorised me to travel with them in the field:

- Beulah Davis, Director of the Natal Shark Board in Durban;
- Professor Leonard J.V. Compagno, of the Grahamstown Institute of Ichthyology;
- Dr Malcolm Smale, of Port Elizabeth Museum.

INTRODUCTION

AMONG THE NUMEROUS activities peculiar to animals, none has ever held our attention so profoundly as predation. Interest in this activity intensifies when man is removed from his customary role of predator to that of prey, especially if this occurs in an environment which is not his own and particularly if it involves a marine aggressor. Even more than the crocodile or the piranha, and no doubt largely because of media sensationalism, it is the shark whose very name conjures up the image of an infernal monster, guided solely by primary instincts and an insatiable appetite. Not only is such a perception inaccurate, but its tenacity still hinders some important scientific investigation. Only in the last few years have subjective speculations begun to give way to objective facts, and it is the aim of this book to take stock of both, at a time when man is investing more and more in the oceans, while the number of great marine predators is not declining.

Even though as a matter of necessity tropical islanders have for a long time watched the habits of the sharks and dogfish around them, in the civilised world interest in sharks was always minimal, even in the first three and a half decades of the 20th century. Not only were interactions between man and shark only sporadically and rarely reported, but few media existed to spread the news. Such ignorance remained the general norm apart from a single group of individuals: coastal fishermen. Their interest in sharks lay as much in the reduction in catches caused by these big predators, as in the reasonable price offered in certain regions for their capture. A few scientists at the American, English or Australian fisheries departments thus made some initial estimates as to the habits of sharks, and confirmed their importance as high-performance predators within the marine ecosystems at the same time as ascertaining their vulnerability to human exploitation.

An understanding of the behaviour of sharks really took off in 1958 after the formation of the Shark Research Panel in the United States, at the instigation of the American Institute of Biological Sciences (AIBS) and the US Navy. This institution was dedicated to involving itself in all aspects of the biology of the elasmobranch fishes (including sharks) and implicitly in the potential danger represented by sharks. This menace was known only in certain regions (eastern Australia, South Africa) until the Second World War, and it was only with the sudden frequenting of all the oceans because of the war, that the problem imposed itself upon all combatants. It persisted, despite the ending of hostilities, with the growing popularity of the sea for sporting and leisure activities.

The idea of reducing the "shark hazard" started to come to the fore with an inquiry into the means of anticipating and controlling the behaviour of sharks. The means available, however, did not at that time permit shark behaviour to be studied in the laboratory any more than in their natural environment. Therefore the scientists directed themselves as a matter of priority towards studying the sensory systems, which seemed to govern the behaviour of these prehistoric predators. In this way they discovered, within about twenty years, extremely sophisticated sensory organs, for the most part unique in the animal kingdom. There remained, none the less, a great general ignorance of the biology of the shark, and, as late as 1985, Dr Gregor Cailliet (Marine Laboratories of California) was still unable to know for sure what quantity of food the shark requires, the frequency of its food intake, how fast it grows, for how long it rests, exactly where it lives, how frequently it reproduces, what its approximate numbers may be on this or that coast, etc.

In the same year, in California, at the time of a symposium on the biology of the Great White Shark, the participants agreed to record a significant increase in the population of this big predator, in proportion to that of its favoured prey: the seal. The report was accompanied by a warning against attacks, which would be sure to increase concurrently in south California. If, moreover, the extraordinary growth in water-sports in the United States, as in all other parts of the tropical belt, is taken into account, it is evident that the number of attacks will continue to rise in absolute terms, even if it remains "acceptable" in relative terms compared with other much less marginal risks of modern life.

As well as these aggressive habits, scientific study of the shark is now beginning to reveal a fascinating and much less uncouth animal than we usually imagine. Its lack of intelligence is amply compensated for by an extraordinary physiology, while on the ecological level its eradication would be even more detrimental to the marine biotope than that of the whales. It is for man to adapt himself to the shark, and not the reverse.

1
STATISTICS
VERSUS MYTHS

TOTAL NUMBERS

- 354 species of shark have been listed by FAO (Food and Agricultural Organization), ranging in length from 15 centimetres (*Squaliolus laticauda*) to 15 metres (the harmless Whale Shark).

- 35 are known to have attacked man at least once, and a dozen habitually do so.

- The biggest Great White Shark ever caught was a female of 6.4 metres, which was caught off Cuba in 1945. It weighed 3312 kilos and its girth reached 4.5 metres; several lorries were needed to tow and transport it. A 10 metre specimen is also said to have been captured, but neither photo nor expert account exists, only subjective statements. The record specimens are always female, since in the sharks, females have larger measurements than males.

- The power of the shark's jaws is phenomenal; the highest recorded with a specially designed apparatus was 3.75 tonnes per cm^2 for a shark of 3 metres. That of a 6 metre Great White Shark must be much more spectacular, but it has never been possible to measure it.

- The fastest shark is the Mako, which can reach 50 km/h (the fastest of all fish is the sailfish, at 113 km/h, followed by the swordfish at 95 km/h). Man's speed in the water is farcical, even compared with the slowest-swimming fish.

- The growth rate of sharks varies according to species, age and maturity: from 33 millimetres per year to 30 centimetres; generally speaking, it is always slow.

- The record for attacks in a single place is at Durban (South Africa), in 1957, with 7 attacks (5 of which resulted in death) within 107 days.

- In August 1960, when a boat capsized at the mouth of the Komati, on the coast of Mozambique, a school of sharks caused carnage among the survivors, mutilating 46 out of the 49 people.

- The speed of tooth replacement in the flesh-eating shark is of the order of 7 to 8 days for the smallest species and from 6 to 12 months for the largest ones, but their replacement is immediate at the time of an attack.

- The shark can smell blood in dilutions of the order of 1 part to 500 million parts of water, and extracts of flesh of the grouper fish of the order of 1 part to 10,000 million.

- In the case of certain large sharks, each seminal vesicle can contain 20 to 25 litres of seminal fluid, or 40 to 50 litres of sperm for a large male.

- The interval between insemination and the emergence of the young shark is 10 to 22 months (one of the longest gestation periods in the animal world).

- All the sharks are capable of going at least 6 weeks without eating; the record observed in an aquarium is 15 months (by one of the species known as "swell sharks").

- Man's most feared predator none the less constitutes one of his food sources: 306,125 tonnes, or 10,000 semi-trailers, was the tonnage of sharks caught for food purposes in 1984.

- Fishing is one of the means of studying sharks, and a means of tagging them on the dorsal fin. In this way, in California, 60,000 sharks (of 47 different species) were captured and tagged in a 23 year period. Nineteen hundred were found again, including one 6,500 kilometres from California and another 19 years later.

- The biggest fish ever caught on a sport-fishing line was a Great White Shark of 1221 kilos, off the Australian coast, on a 60-kilo line (source IGFA: International Game Fish Association) – the biggest ones broke the lines.

- The oldest ancestor of the shark was discovered as a fossil, the "cladoselache", 350 million years old – this is 100 million years before the dinosaurs appeared; man first appeared 3.5 million years ago.

PRECONCEIVED IDEAS

Before going in detail into all the scientific studies or the indisputable evidence which have permitted a much better understanding of the world of the "Squaliformes", it would seem of interest to put right certain falsehoods, approximations, even scientific blunders, concerning these fascinating animals.

- Sharks attack in order to feed. This image of the insatiable shark that swallows everything it encounters in order to assuage a hunger that is never satisfied is one of the most false. According to Balbridge and his reference work on 1500 cases of attacks on man, 50-75% of these attacks have nothing to do with nutrition. This does not resolve much concerning the danger linked with sharks, but we shall see that it is a determinant parameter for explaining, on the one hand, certain single bites, which are not repeated and do not involve the removal of flesh, and, on the other, the effectiveness of certain preventive measures.

- Hunger does not exist in sharks. In apparent contradiction with the preceding proposition, this is none the less untrue. The shark is capable of very long periods of fasting, and it seems that it may have a periodicity in its eating habits. During these "eating phases" it is evident that we can talk of "hunger", and attacks will then be different in their determination and development.

- The shark is a primitive animal. Even if it is true that the order of fishes is the most rudimentary in the hierarchy of animal evolution, the shark is much more sophisticated than we believed it to be only 20 years ago. Besides sense organs that are particularly outstanding and unique in the animal world, the size of its brain is closer to that of birds (proportionately speaking) or of certain primitive mammals than to that of any other fish. We shall see for instance the exceedingly sensitive character of the olfactory organs in the shark, reflected at brain level by highly developed olfactory lobes. Several other physiological characteristics prohibit any talk of a primitive animal and, moreover, such a creature would not have come through 350 million years without appreciable evolution.

- The shark has very poor eyesight. Even Captain Cousteau subscribed to this unfounded belief. Not only can it see contrasts well, but it has good night vision thanks to a histological structure peculiar to nocturnal animals (the *tapetum lucidum*). It can even see colours.

- The shark's pupil is at all times widely dilated, adding to its terrifying appearance. This is false: the iris has an ovular shape and the pupil can be dilated and constricted very rapidly. Some species possess a third eyelid (nictitating membrane) which closes at the moment of attack (as in the case of all the Carcharhiniformes, including Tiger, Bull and Oceanic White-tip Sharks).

- The shark attacks to defend its territory. This has never been demonstrated for any species, even for the reef sharks which are fairly sedentary. This concept of territory is standard for terrestrial animals which set the boundaries of it by their droppings, but the extension of the principle to fish seems without foundation. Perhaps we could speak instead of defence of a vital space immediately around the shark when it feels threatened.

- The shark rolls on to its side to attack. The shark's anatomy and scientific observation refute this popular belief, which must date back to ancient times, as Pliny mentioned it. The shark approaches its prey from an angle, plants its pointed lower teeth into the flesh, then brings its upper jaw forward to plant its cutting teeth in.

- The shark does not eat dead bodies. Ten or so examples will be cited in the book which categorically refute this belief, some of them supported by photographs. In 1950, Mr Warne, an Australian fisherman, found the barely digested right hand of a human being in the stomach of a 1.5 metre Tiger Shark. The police identified it as belonging to Peter Szot, whose body – minus the right hand – had been recovered on a beach eight days earlier, with a bullet in the head. Mr Szot had not committed suicide after having lost his hand, but before.

Preconceived ideas are not very important when not put down in writing. When, on the other hand, they claim to be guiding rules for the attention of the trusting reader, they can lead to a lethal ignorance. This is how we came to read the following absurdities in the survival manual of the US Army, published at the start of the last war:

"The shark is a cowardly fish which moves about slowly, easily frightened by surprises in the water, noise, movement and unusual shapes. This last point alone would be enough for a shark not to attack man." This cartoon-character description was followed by a paragraph on the best way of combating this "wretched" fish: "First of all strike it on its soft and vulnerable snout or in the eyes, or stab it in the gills." The authors then boldly advised "swim outside the line of charge, catch hold of a pectoral fin as it passes, and ride with it for as long as you can hold your breath." The most savoury piece of sitting-room advice imaginable, but this was only a start. "If you can manage to cling to it, the shark may lose some of its vice and regain its natural cowardice. If you have a knife, open up its stomach. On opening the stomach, you cause water to enter – that will kill it almost instantaneously." Phenomenally stupid words, in which the US Army pilots and marines were supposed to immerse themselves before their missions.

One reader of the manual, who managed to survive a shark attack in spite of the book's advice, related how he fired his Colt 45 at the "soft and vulnerable snout". "It then turned around to charge, and so I started to bang it on the top of the skull with the Colt. Its head was as hard as metal at this spot and I later discovered that I had partially flattened the small steel eye on the grip, at the point where the strap is attached."

The present-day "survival book" of the American army is reduced to the basics, as if the authorities had opted for a helicopter behind every GI, rather than for a good manual in every battledress.

2

AN HISTORICAL ACCOUNT

THE FIRST REPRESENTATION of an attack by what must have been a shark was discovered on a vase unearthed at Ischia, Italy, an island just west of modern day Naples. The vase shows a man seized by a fish reminiscent perhaps of a shark, and has been dated *c.* 725 BC.

The first account of an attack by a marine monster dates back to Greek history, with Herodotus in 492 BC. He was not talking specifically of the shark, even though the latter was probably involved, for the word did not yet exist and no really lifelike graphic representation was to appear before the 18th century. Again in Greek history, the poet Leonidas of Tarentum evokes the tragic end of the sponge-fisher Tharsys, when he was being hoisted aboard his boat by his two companions and was attacked by a sea monster which tore away the lower sections of his body. Tharsys' companions brought ashore his remains and thus, the poet elegantly noted, Tharsys was buried both on land and at sea.

The first reference in English dates back to 1580 when an officer related an attack he had witnessed between Portugal and India. "A

One of the first representations of a shark attack, by the Swede Olaus Magnus around 1550

Medieval representation of sharks

man fell overboard during a storm, and it was impossible for us to reach him or go to his assistance in any way. So we threw him a block of wood attached to a rope, specially provided for this purpose. Our crew began to bring in the man, who had managed to catch the block, but, when he was no more than half the range of a musket away, there appeared from beneath the surface a big monster known as tiburon; it rushed at the man and cut him to pieces right before our eyes. It was certainly a terrible death."

In 1776, Pennant described the Great White Shark: "They reach very great dimensions. There is a report of a whole human corpse being found in the stomach of one of these monsters, which is by no means beyond belief considering their huge fondness for human flesh. They are the nightmare of seamen in all the hot climates, where they constantly follow ships waiting for anything that might fall overboard. A man who has this misfortune inexorably perishes. They have been seen to rush at him like a gudgeon at a worm... Very often, swimmers are killed by them. Sometimes, they lose an arm or a leg, and at other times are cut in two by this insatiable animal."

Ships' logs often recount similar tragedies, but a few rare cases exist in which the seamen get the better of the situation. The captain of the Ayrshire fell overboard in the course of a crossing in 1850. His courageous labrador plunged into the water to rescue him. A shark immediately headed towards them, but, according to the logbook, both were saved. The captain was unharmed, but the dog's tail had been cut clean through.

Basic scientific knowledge of the shark is extremely recent, and it can therefore be assumed that up to the 20th century, accounts of attacks by sharks were largely based on popular mystique, superstition and fanciful speculation.

DEVIL SHARKS AND GOD SHARKS

The Greeks wrote their legends from the constellations, but well before them primitive humans projected various representations of their own devil-god, the shark, on the stars. The stars that the Greeks saw as the belt of Orion were for the Warran Indians of South America the missing leg of Nohi-Abassi, a man who had got rid of his mother-in-law by training a murderous shark to devour her. As legions of men have discovered since then, Nohi-Abassi learned to his cost that it is not safe to provoke a shark or a mother-in-law. His leg was cut off by his sister-in-law, apparently playing the part of the shark, and Nohi-Abassi died. His leg wandered into one region of the skies, and the rest of his body into another. For some primitive tribes the shark was an avenging god, for others a two-faced devil. In many primitive religions, the status of the shark became so complex that it had several roles: sharks became men, men became sharks. On many Pacific islands, the insatiable god could not be satisfied by the men, women or children which it occasionally gulped down in the depths of the sea – so it claimed the ultimate homage: human sacrifice. The head of the high priests then made his way among the people accompanied by an assistant, wearing a nose similar to the long snout of the shark. At a signal from his chief, the assistant pointed his nose towards the crowd. The person, man, woman or child, at whom the nose happened to be aimed was immediately seized and strangled. His body was ritually cut up into pieces and thrown into the sea for the shark-gods.

In the Solomon Islands, the deified sharks lived in sacred caves constructed for them near the coast. Opposite these caves, large stone altars were erected, on which the bodies of the chosen victims were placed. After mystical ceremonies had been performed, the bodies were offered to the sharks. Certain sharks in the Solomon Islands were considered to be incarnations of deceased ancestors; these were the good sharks. Other estranged sharks, which roamed between the islands on fiendish missions, were considered malevolent. The fishermen could, however, drive out these evil-minded sharks by brandishing in front of them small wooden statuettes representing the familiar sharks.

All these appalling rites and traditions still existed only a few decades ago in certain isolated islands, and the more "civilised" among them still persist.

The Vietnamese fishermen still refer to the Whale Shark in its capacity as Ca Ong or "Mister Fish". Little altars beseeching the protection of Ca Ong can be seen on sand dunes all along the central and southern Vietnamese coast, close to wrecked tanks and other relics of the war.

When the US Navy built the enormous base at Pearl Harbour in Hawaii, the remains of an enclosure where the Hawaiian kings used to make local gladiators fight with captive sharks were found. The sharks represented the ancestors, and were fed with live humans. The only weapon the gladiatorial warrior could use in his defence was a single shark's tooth mounted on a short wooden handle held tightly in the fist. Only the tooth showed between the two fingers firmly closed around the handle. The difference between this and the matador, who can dodge a charge by the bull, was that the warrior was only allowed one chance and that, to win, he had to let the shark charge him. He was supposed to wait until the last moment before diving beneath his assailant and trying to disembowel it with his weapon. Legend has it that sometimes, on rare occasions, the warrior got the better and killed the shark. Perhaps a royal edict stipulated that if the warrior drew blood from his adversary he could leave the infernal arena, failing which it seems impossible that such duels could have been terminated in any way other than in victory by the shark, which after all did have a mass of teeth to set against the single one held in the fist of its opponent, to say nothing of its swiftness of manoeuvring in the water. The shark arena was a circular enclosure of about one hectare, made up of lava rocks. It had an opening on the seaward side to allow water to enter. Fish and human bait were thrown into the enclosure in order to attract sharks into the opening, which was of course shut at the time of the "battles". The queen of the sharks was supposed to live next to the arena, at the bottom of the bay. The queen condescended to permit battles near her refuge, provided that she was seduced by offerings. These offerings were once again human, for one of the economic realities of life in Hawaii in former times was that people cost less than pigs.

In 1900, when the US Navy had completed the construction of an enormous dock at Pearl Harbour, at a cost of four million dollars, the foundations suddenly collapsed under the pressure of an under-sea eruption, and the whole dock sank under the waters. The engineers looked into all the hypotheses to no avail, but the natives knew what had happened: "The queen of the sharks is angry and flexing her back." Even today, numerous beliefs persist with regard to sharks, not only in Hawaii but also in Tahiti, the Cook Islands in the Torres Strait, the Marshall Islands and Samoa, and even among the Alaskan Indians and in Latin America, where many ancient pieces of pottery have been unearthed depicting swimmers being devoured by sharks.

In Japan, one of the mythological gods is the storm god, known as the "shark-man". In fact, the shark is so terrifying in Japanese legends that when the Chinese thought about a talisman to paint on their aircraft to fight the Japanese, they chose the demon head of the Tiger Shark. The American pilots who did the same were known throughout the world as the "Flying Tigers", when in fact they should have been called the "flying sharks".

In the Torres Strait, between Papua and Australia, Mutuk is a legendary man who was delivered from the stomach of the shark which had swallowed him. His "gastric adventure" naturally brings to mind Jonah, who was swallowed by a "big fish". As the biblical scribes were unlikely to distinguish between a fish and a mammal, it was thought that a whale was probably involved, and most religious representations of the event stick to that animal. Anatomically, however, it is difficult to see how a prophet could have passed

During the colonial period, sailors learned how to catch sharks. The capture was always a major event on board, and even Napoleon was invited to the spectacle on board the *Bellerophon* after his defeat in 1815.

through the whalebone of a whale, and so it must indeed have been a big fish, and probably the biggest of all fish, the shark. Moreover, this shark could not have had a mouth filter like the Whale Shark, but it would have had to have been large enough to swallow a man whole. Therefore it could only have been the Great White Shark (*Carcharodon carcharias*). And let readers of the bible be reassured, whale or shark, the regurgitation of a live man is even more miraculous from the inside of a Great White Shark than from a whale.

If sharks are legitimately the origin of many myths, they can also, in certain circumstances, be a brutal means of debunking. In 1776, the naturalist Thomas Pennant wrote: "The master of a slave ship from Guinea told me that a wave of suicides had taken hold of the recently bought slaves, for the poor wretches thought that after their death their bodies had to be returned to their family, their friends and their country. To convince them that their bodies would *not* be condemned to perpetual wandering, he ordered that one of the slaves be tied by the ankles to a rope and lowered into the sea. He had scarcely been underwater a minute when the crew pulled up the body, but only the ankles and the feet protected by the rope remained intact. All the rest had been devoured by sharks." Thus it was bluntly demonstrated that the mortal remains would not have to be returned to the families since they would be fed to the sharks.

If India has its snake-charmers, the Fiji Islands have their shark-charmers. Twice a year, to avert the sharks from attacking them, the Fijians would indulge in the "ceremony of kissing the shark". Father Laplante was a missionary in these islands up to 1938 and told how the sharks were captured in a large net, turned over on to their backs by the slightly drugged officiating ministers, before being kissed on the stomach. The missionary was astonished that on each occasion the sharks, once kissed, stopped moving, "as if the men had an occult power that I wouldn't know how to define". This custom experienced a renewal of interest in 1960, at Fort Lauderdale in Florida, when students made it an initiation rite for new pupils. The police put an end to this very unusual ragging by putting a close guard on the shark they named "Freddy", before returning it to the sea. Although it measured only 1.5 metres, it was a Tiger Shark.

The pearl-fishers of Ceylon likewise resorted to shark-charmers to protect them. The powers of these charmers were hereditary and mystical in the extreme, as Sir Emerson Tennent reported in 1861, since even if the charmer was ill and incapable of moving about, it was sufficient for him to delegate anybody in his place for the sharks to remain placid. Without wishing to deny this highly mythical power, I shall say that the same officiants would certainly not have had the same results on the Great Barrier Reef of Australia. We shall in fact see that, curiously, and despite its latitude in the Indian Ocean, Sri Lanka has always remained relatively safe from attacks.

The mystique attached to sharks has managed to make a fortune for some, if we believe François Poli's book *Les requins sont capturés la nuit* (1959, Chicago). Lake Nicaragua is well known in Central America for the numerous "man-eaters" it harbours, and the book concerns an event that prompted the resident lakeside Indians to attempt to appease the "lord of the waters". They had very elaborate funeral ceremonies, at the end of which the bodies, covered in jewels and weighed down with gold ornaments, were committed to the sharks of the lake in order to appease them. The latter of course devoured the bodies together with the jewels, to everyone's satisfaction. Until the day when a Dutchman interfered by hunting the sharks, opening their bellies and retrieving the gold and the sacred jewels. He very quickly amassed a fortune, but forgot to leave in time. Informed of his sacrilegious thefts, the Indians set fire to his house after having slit his throat. His corpse being unworthy of the sharks, it had to be carbonised.

Certain myths have given rise to real events, like that we have just mentioned, and certain events could have given rise to myths had their improbability not been far too gross. For instance the story told by Marc Twain, and regarded as a true fact for some years. According to him, a certain Cecil Rhodes was said to have caught, close to Australia, a shark which had swallowed a newspaper in London ten days previously; the lucky fisherman thus learned before everybody else that the wool market had shot up considerably and he invested considerable sums risk- free, which turned out to be the beginnings of an enormous fortune.

Certain shark attributes, alleged but not proven, easily become superstitious beliefs. For example, the ability that sharks are supposed to have to "smell death". Many seamen who died aboard ships and whose bodies were committed to the sea in fact did find their graves in a shark's stomach. But the superstition grew with time until sharks were believed to be able to know when a man was on the point of dying, and the appearance of a shark in the wake of a ship became a sign of imminent death on board.

When a cholera or yellow fever epidemic broke out on board a ship, the superstitious believed that the sharks would remain behind them until the epidemic had delivered its final victim. A skipper who sailed from San Francisco added to the legend. He often carried an unusual cargo: the corpses of Chinese people who had died in the United States, and who, according to ancient custom, had to be

This illustration from the 19th century represents the shark as an exotic curiosity rather than a "killing machine"

buried in China. This skipper was categorical: when transporting bodies, his boat was followed by an army of sharks as if they were able to detect the corpses encased in plated coffins down in the hold. The sharks never appeared when his cargo was less macabre.

It is curious to note that in the brotherhood of seamen, so respectful of traditions and even of superstitions, there has never been a prohibition on naming a ship *Shark, Tiburon* or *Requin*. When one knows the price to be paid for uttering the mere name of the "animal with big ears" (the rabbit) on board a ship, it is surprising that the shark, which has nevertheless terrified generations of seamen and castaways, is not put in the same category. In the United States alone, six ships have been named Shark: the first, in 1821, was a twelve-gun schooner, and the other five were submarines, the last of which, launched in 1960, was a nuclear-powered submersible.

FROM MYTH TO SYMBOL

The etymological roots of the word "shark" in fact indicate certain characteristics of the animal itself. The Anglo- Saxon root *scheron* means to cut or tear (compare the French *arracher*). "Schurke" is the German word for villain. Since Elizabethan times, popular speech has used the word to indicate "sharks" in various contexts: loan shark, pool shark, card shark etc. The sound itself is sharp, and emphasises the impression of urgency, of terror, of surprise and of assertiveness.

In France, the lord of the seas goes by the name "requin". This very probably comes from "requiem", evidently referring to the unenviable fate that awaited the unfortunate seaman falling overboard in certain tropical waters. In Spanish, the animal is referred to by the name "tiburon". Not surprisingly, perhaps, many people in these countries believe that the English name for shark is in fact "jaws", from the celebrated top-billing film.

If one asks a random selection of townspeople and professional shark fishermen to choose four images that immediately come to their mind when the word "shark" is mentioned, the similarity of response is amazing.

"Danger", "killer", "man-eater", "jaws", "desperate", "teeth", "fin", "fishing", "Great White", "Mako", "marauder", "foam", "deep water" are some of the key words most frequently encountered, irrespective of occupation, of the country concerned and of any possible real-life experience of the person in question with regard to sharks. It is remarkable, in fact, that all of the people questioned have never witnessed an attack, some have never seen or even caught a glimpse of a shark in their lives, and others live in countries where there are no sharks.

These responses suggest that the image of terror and destruction is much more firmly established than the picture emerging from scientific discoveries which reveals the shark to be an animal with a very distinctive biology and behavioural characteristics. Several factors may explain the shark's reputation as a symbol of power and terror.

- It is undeniable that the shark really is the "lord of the seas" and the ruler of the deep of early ages, since it has existed unchanged for 350 million years, while man did not even exist 1 million years ago. The shark is not only the biggest of the fish, it is also the best equipped for hunting down and destroying its prey.

- It is capable of living in all waters shallow or deep, tropical or temperate, fresh or salt. And as it is perpetually moving, there is nowhere that man can feel really safe from its attacks.

- Unlike most animals, the shark is never in need of prey on which to feed, and it is even capable of devouring its own congeners. It has no natural predators apart from the Killer Whale and, very occasionally, the Sawfish. Its "invincibility" is physical, dietary and ethological.

- Even more so than other fish, its tenacity for staying alive is impressive. Gaffed, shot, harpooned, ripped open, it is still capable of moving about and tearing apart its victims in the water. Even when apparently lifeless on the deck of a boat, it can still seize and cut off the arm or the leg of the imprudent fisherman several hours after being captured.

THE REASONS FOR FASCINATION

Anyone investigating the man-shark relationship will be struck by the disproportion between the impact on the subconscious of a few exceptional mishaps and the statistical reality of the facts. The recently established organisation International Shark Attack File, which lists all known attacks since 1560, was unable to find more than 1500. Even though the true number is, in my reckoning, much higher, it is nevertheless the case that the number of attacks throughout the world certainly does not exceed 500 per year and of these only 200 will die (if we disregard big shipwrecks). This is the number of people struck by lightning each year, the consequences of a single reckless weekend on the roads of France, the number of victims in a single air disaster, the number of people dying of AIDS every two weeks in the United States. So why this irrational fear? Why all the media coverage given to these deaths, which are then talked about in minute detail the world over?

At the root of it there is certainly a phenomenon of projection, that unconscious action which consists of refusing to see or being incapable of interiorising certain emotions or certain personality traits, and instead "pinning" them on another person, another place, another object. The classic example is that of the "bad element" which always seems to linger in the great families and drain the guilty conscience of one and all like an abscess. Dogs, meanwhile, are often an outlet for their master's frustrations when the latter insults them as he would like to dare insult his boss or mother-in-law.

We can each one of us ask ourselves if there are any aspects of our own personality that we project on to the shark or, on the other hand, any shark characteristics which we would secretly like to be able to have at our disposal. In a later chapter we shall see the thousand and one ways of making use of sharks, introspection no doubt not being the least of these.

We have seen how island peoples have always attributed both attractive and repulsive qualities to sharks, as if primitive peoples, living in close contact with the earth and the natural cycles, more easily saw both the negative and the positive side in everything. It is significant that nations called "civilised", which have lost touch with nature, project onto the shark an almost exclusively morbid, negative image. The gap is ever wider between our safety-conscious, hypersecure, ever more comfortable and technologised world, and that abyssal, cold and sombre world which is that of the shark. A world which takes us back to the barbarity of early ages when the frail body of man did not count for much faced with physical assaults of every order. It is probable that the shark will convey more and more frightening images as modern man creates for himself an artificial world which is more and more fearful and less and less "physical", and in which the "hunting instinct" will have disappeared for good. The "organic" relationship which we maintain with the shark will become increasingly intolerable, and at the same time will be increasingly exploited by the media, as if the better to exorcise these hideous reminders of another age.

The fear of sharks embraces a number of phobias: fear of the dark, of heights, of falling into a strange world, of blood, of gaping wounds, of being alone when faced with danger, of physical combat, of the inexorability of death... These chilly monsters, which glide through the water like serpents and lock you in the stranglehold of their jaws as in the clutches of a spider, take you back to the nightmares of your childhood.

We must not, of course, overlook the "delightful shudder of horror" as a possible reason behind this fascination for the shark. Sadomasochistic drives still exist in many adults, and whether these drives are active or passive, the subconscious still projects onto the shark or its victim. The morbid instinct of many readers is well known among editors-in-chief, who know how to distil the horror of an attack across eye-catching headlines, their tales of suspense and photos often having nothing to do with the actual event. Do you need proof? Watch the attitude and reactions of a potential reader of this book when confronted by the sealed photo section it contains. And then, how did you yourself react in the bookshop when you came to these photos that were hidden from you?

Our fascination may become more admiring, if not wholly positive, when we come to know more about the invulnerability and hyperefficiency of this unpredictable, uncompromising and solitary animal, which trifles with others as it pleases. Our interest may then turn more towards the animal than towards its victim.

Whatever the reason behind man's current fascination for sharks, it is always accompanied by anthropomorphism, inseparable from the phenomenon of projection, which is encountered every time man is confronted with a problem outside his control. There was the serpent, there was the bear, there will soon only be the shark left, as long as it retains some of its mystery. Having arrived on earth 350 million years before man, it will without any doubt outlive him, and the passage of man in the animal world will have been only a minor incident in the history of the shark.

SHARKS AND THE MEDIA

The shark is a creature tailor-made for media sensationalism. Magnificently represented by the Great White, it is a primitive carnivore of terrifying appearance, whose diet does not exclude the odd human being now and then. In other words sharks possess all the necessary credentials for a front-page news story. However, as the latter is normally reserved for events concerning people, sharks do not become newsworthy until they have just devoured a potential reader.

Although journalists are experts at collecting data, they are not shark specialists, and in general they have very little available time to write their article. This explains the scientific impoverishment that most often presides over the reports of such events, and the stereotypical manner in which accidents are described. Contrary to these reports the circumstances are usually totally different, sharks are not always "ruthless Great White Sharks", the victims not necessarily "young, athletic, swimming champions" and the waters not automatically "calm and blue with nothing to warn of the drama about to unfold".

Besides attacks, there should be many other foci of interest for journalists writing about sharks, such as the way of fishing for them, cooking them, detecting them, avoiding them, repelling them and luring them with bait – but the stories would no longer be front-page material, indeed they would barely merit a few lines in the "tourism" column for the attention of those going on a "game-fishing" trip to Mauritius in December.

When the celebrated film *Jaws* was released at the end of the 1970s, it was a potential goldmine for all the critics, who could have made not only a cinematographic but also a scientific analysis of it. The technical adviser of this film was Professor Compagno, without doubt the greatest specialist worldwide, and, as he himself told me, the implausibility of the film came not from the attack sequences but from the exaggerated accumulation of them and from the excessive anthropomorphism with respect to the shark.

Our modern society is in the end less well equipped to separate myth from facts with regard to a film like *Jaws* than was the society of the 19th century with regard to the novel *Moby Dick*. Whaling was a standard industry at the time when *Moby Dick* was written, and everyone knew that whales could kill their hunters. Most of the inhabitants of the New England coasts knew personally men who had been whaling or had at least seen whales, and they knew that the "Cachalots", or Sperm Whales, do not devour a man deliberately.

But we are no longer close to nature, for it is no longer indispensable to our survival, and our gullibility is all the greater. The media could have a dual role of informing and educating via a film such as *Jaws*; and as a result the reader's or the spectator's interest would be greater still, and the role of the journalist might turn out to be more gratifying. Many people have seen the sequels, *Jaws* 1, 2, 3, and 4, but who has dissected for them the alarming implausibility of the scenarios and the unreality of the behaviour of the new heroes?

In the countries where such things take place, many journalists dream of covering a shark attack. One of them even left a town in the centre of Oregon for another on the coast, just to be first to cover the next attack. It is true that any shark attack will always be a newsworthy event as long as it remains exceptional. Imagine an American reading on page five of his newspaper of 30th June: "About 250 people will die during this first summer weekend on all the nation's beaches, through attacks by sharks. This figure represents a very clear improvement on the 375 attacks in 1990, before the speed limit of 55km/h prevented a considerable crowd from getting to the beaches more quickly...".

This type of article appears every year in connection with road accidents, and is received with general indifference, despite the tens of thousands of deaths, to which we have become immune. And then, what could be less mysterious than a car accident the circumstances and causes of which are always known, the speed of which allows no time for states of mind or for suspense, the people involved in which belong to the same world, and the reporting of which leaves hardly any room for interpretation?

The journalist Steve Boyer tells a story indicative of the ever greater extravagances to which anecdotes relating to the world of sharks lend themselves. A few years ago, the news agency Associated Press received the following dispatch: "Is the population boom in seals attracting more sharks?". The question, a banal one, came from a small local newspaper on the west coast, the Santa Barbara News Press. The message was taken up by another paper, this one regional: "More seals may mean more sharks". The next day, another printed on its front page: "Experts declare that the seal population is attracting more and more Great White Sharks... A debate on the causes of the proliferation of the Great White". A few days after the initial dispatch, another paper announced across five columns: "Great White Sharks infest Santa Barbara waters". We are a long way from the original dispatch by a small local editor, who certainly did not imagine that because of him the tourist trade at his seaside resort would collapse for several months... Not content to follow the generalised attempts to outdo one another, a final newspaper sank into total approximation: "Sharks and seals attacking divers..." In the journalists' profession, as in many others, everything depends on the ethics one has set oneself: whether to give way to sensationalism or not, whether or not to sacrifice the reality of the facts for the extraordinary stories the public clamours for. It is significant that the very responsible *New York Times* prints 900,000 copies daily, whereas the newspaper specialising in sordid murders and other small news items, the *New York Daily News*, is probably all set to do a special issue on the minor threat of shark attacks in the Hudson, and prints 1.5 million copies.

Much more innocuous is the anecdote related by that great specialist David H. Davies. In August 1959, from the bank of the mouth of the River Umgeni near Durban, two fishermen captured an impressive Bull Shark 1.8 metres long and weighing 120 kilos. After having fastened it by the tail to a stout tree stump in a few centimetres of water, they left to inform the people in charge at the Durban aquarium of their catch. The latter were very interested in such a species, known as the most dangerous in the region, and a lorry was sent to bring the shark back.

It was a lifeless body covered with a damp sheet that arrived several hours later at the aquarium, where it was immersed in a tank for observation. To everyone's surprise, a few minutes later the shark was swimming vigorously around the tank, and four remoras had already immediately adopted it as their new "master". This was by far the most spectacular of the fish that had ever stayed at the place, and from the very next day a crowd of visitors began to flock there. For twenty-three days the shark adapted remarkably well, but did not eat. It was only when attempts were made to save a young manta ray of 5 kilos, which had been salvaged from a fishing net and placed dying in the aquarium, that in a cowardly fashion, the shark attacked and swallowed it in one second.

It very quickly became the "boss" of the aquarium, and its popularity spread. A local newspaper named it Willie, and the name was to stay with him for good. Willie fed irregularly on the pieces of shark or ray flesh that were thrown into the tank but preferred to manage by himself. Throughout the period from September to December, several rare specimens disappeared down Willie's throat, with the exception of four Dusky Sharks and four or five dogfish including one pregnant female. However, the latter was soon cut in two by Willie, just before her several young were to be born. And then not content with spreading devastation among the animal life around him, from December onwards Willie seemed to take an interest in the divers who regularly carried out routine work in the aquarium. This perhaps had something to do with the rise in water temperature, but knowing the past record of this type of shark in relation to human beings, those in charge were not about to take a risk. The situation was not a simple one, for Willie had become the visitors' most favourite resident by far and the removal of such a specimen would not have been popular. Moreover, if the animal were set free, no doubt all subsequent attacks in the region for years to come were going to be attributed to him. There was also the not inconsiderable problem of how to capture him inside the tank.

After much serious thought, it was decided, very reluctantly, that Willie *had* to be captured and removed. At dawn, for three mornings in succession, several methods of capture proved ineffective. In the end, Willie was brought to the surface with a large triple hook attached to a heavy nylon rope and immediately killed and cut up into pieces which were concealed in several dustbins. By seven in the morning everything was completed, there had been no witnesses to the "murder".

In the half-hour following the opening of the doors to the public, a reporter rushed up, having been informed by a visitor that Willie had disappeared and wanting to gather any information on the event. David Davies, the establishment's Director, decided not to launch forth into details, and revealed that, in actual fact, to everybody's great sadness and utter surprise, Willie had been found dead beneath the surface that same morning, at dawn. The reporter piled on the questions, becoming more and more inquisitive, and finally asked whether an autopsy had at least been ordered. Considering that the ignominious cutting up that had been carried out earlier that morning could be regarded as a post mortem examination, Davies replied that, in fact, the autopsy had indeed been performed. The next question demanded what this had shown, and, as Davies had certainly noticed a slight and commonplace discolouration of one of the lobes of the liver, he pointed this out to the reporter, who, to everyone's relief, finally agreed to leave. The next day, across five columns on the front page, the Durban local printed: "Willie dies suddenly from liver failure".

A CHAPTER OF HISTORY

It was not until 1916 that stories about sharks were to leave the realm of myth and legend to which they had hitherto been confined. For in that year the sensationalists were to have a ball and the morbid to take delight, the stubborn disbelievers were forced to keep silent, the sceptics to become observers, the scientists to ask themselves questions, and the political authorities to wake up. Fanciful speculations would henceforth give way to attempts at objective analysis, even though it would need another sixty years to arrive at the scientific certitudes chronicled in this book.

Saturday 1st July 1916, Beach Haven, New Jersey

Charles Van Sant runs towards the beach, eager to plunge into the cool water he has been dreaming of since the beginning of the week. He has only just got off the train from Philadelphia, accompanied by his father and his two sisters whom he was too impatient to wait for. Within minutes of arriving at the hotel, he was in his swimming costume and running out of the foyer. At twenty-three, he will soon no doubt be dragged into the war like so many others but, for the time being, the endless horizon offers itself to him and he plunges with delight into this sea that he loves so much. The sea is calm that day at Beach Haven.

Charles is a strong swimmer and very soon is far out from the beach, going a hundred metres or so beyond the distant barrier. After a few minutes, he decides to return towards the shore, and turns his back on the open sea, as if with regret, now swimming lazily, unhurried as he is to interrupt this first serene and solitary bathe. But he is no longer alone.

Just behind him, tracing a beeline wake beneath a black fin, a grey shadow is catching up with him. It has been seen from the beach, and bathers shout at the swimmer but he doesn't hear. They suddenly stop, speechless and immobile, paralysed at the sight of the shorter and shorter distance now separating the two silhouettes. Van Sant is still swimming slowly, unable to imagine for a single moment that he might be the target of some deadly pursuit.

It is when he is very close to the shore that the water seethes around him and red foam encircles his body. Immediately, Alexander Ott, a former member of the American Olympic swimming team, dives in and swims faster than he has ever done before. Just as he arrives level with the red stain, the grey shadow turns around menacingly, slowly approaches, then rapidly disappears into the blue waters, leaving Van Sant to the man who has come to rescue him.

Ott manages to bring Charles back to the beach, surrounded by a crowd horrified by the sight of his legs which are cut to shreds. Van Sant dies that evening of haemorrhagic shock.

The grey shadow departed as it had arrived, invisible and mysterious. Nobody could remember a shark ever having killed a swimmer in the past. Perhaps it had happened in the waters of the open south or in Australia, but never in New Jersey. And what about the experts who declared that there had never been any absolutely authenticated case of a shark attacking a swimmer anywhere in the world? Twenty-five years beforehand, a rich New York banker had offered a prize of five hundred dollars to anyone who could prove to him that a swimmer had been attacked by a shark anywhere north of Cape Hatteras. The prize had never been claimed.

Three years earlier, on 26th August 1913, a fisherman had caught a shark off Springlake, in New Jersey. Although a woman's foot with a leather shoe and a stocking had indeed been found in the stomach, this simply appeared to prove that sharks devoured dead bodies, not live swimmers.

6th July 1916, Springlake.

It is five days since Van Sant was killed. Over five hundred people are out for a stroll on the beach. It is low tide, and very few swimmers are in the water. Springlake is an elegant and peaceful resort, frequented by all the upper middle class of New York. Senators and governors live close to the shore there in their luxury houses which they like to call "cottages", or in the big hotels, notably the Essex and the Sussex. The talk is neither of the war nor of that lowly young man who died a few days earlier at a rather antiquated resort 80 kilometres away, but about the epidemic of infant paralysis that has been decimating New York for weeks, with twenty-four deaths on 5th July alone.

In the sea of democracy, a bellboy or page is just as good as a billionaire, and that is perhaps why Charles Bruder loves the ocean. When he is not working at the Essex or the Sussex, he often goes for a bathe in the course of the day, and everyone knows him as one of the faces of the town. He is only twenty-eight, but his open nature and his kindness have made him popular with everyone. With his tips, he maintains his only family: his mother who lives in Switzerland.

Bruder is not working on the afternoon of 6th July and, low tide or not, he is determined to go for a bathe. He walks out almost as far as the barrier, talking to and smiling at the clients who recognise him. When the water is up to his waist, he decides to dive and start swimming; he very soon goes beyond the "security lines" which enclose the bathing zone, but White and Anderson, the lifeguards on duty, do not intervene as they would with the majority of bathers, for everyone here knows that Bruder is an excellent swimmer.

A woman's scream echoes on the beach at Springlake and, instinctively, White and Anderson scrutinise the sea. Bruder has disappeared. "He's turned over!" the woman cries out. "The man in the red canoe has turned over!"

She has scarcely started to scream again before White and Anderson have rushed into their dinghy, heading for what is not a red canoe as the panic-stricken woman thinks, but the cloud of blood in the middle of which emerges the dying face of Bruder and, for a brief instant, one of his arms dripping with blood. The boat reaches him and White offers an oar to Bruder who still has the strength to grasp it. They pull him towards them. His face is terribly pale and his eyes are closed. "A shark, a shark's had me, taken both my legs", he still has the strength to moan before losing consciousness. White hoists him over the freeboard, his body does not weigh much. When White and Anderson arrive at the beach, they hesitate to set Bruder down in the midst of the crowd, where several woman have already fainted. There is in any case nothing more that can be done for him.

The switchboards at the Essex and the Sussex telephone all areas and, in a quarter of an hour, all the swimmers have left the water along the thirty-five kilometres of the New Jersey coast.

But was it a shark? Is it true that man-eaters will now plague the region's coasts? The hoteliers, resort officials and gossip columnists all wish to make it known that it cannot have happened, and people start to talk of a turtle or an enormous mackerel to account for Bruder's wounds. All wait anxiously for the verdict of the doctor, Colonel Schauffler. The latter's decision is final: "There is not the slightest doubt that it is indeed a man-eating shark that has inflicted these injuries on Bruder. The right leg has been torn off, and the bones cut halfway between the knee and the ankle. The left foot has been torn off as well as the lower part of the tibia and of the fibula. The bones are stripped of flesh below the knee, and a deep gash stops in the femur above the knee. On the right side of the abdomen, a piece of flesh the size of a fist is missing."

That night, while a collection is being made for Bruder's mother, motorboats equipped with searchlights are launched for a futile pursuit. The crew members are armed with guns for patrolling, and fishermen set tens of lines with mutton which is reputed to be the best bait. "I am certain that two or three days from now the beaches will be safe", Senator Hill declares. Not one shark is captured, shot or even seen.

The day Bruder is killed, twenty-four people die in New York from poliomyelitis, known at that time as "infantile paralysis", but the newspapers talk only of Bruder. Such are the demands of man's terror and his fascination at the hands of the shark.

In the days following, there is a frenzy of action and unhelpful commentary. At Atlantic City, swimming costumes that do not cover the hands and feet are banned, while at Asbury, with the help of publicity, the installation of a shark-proof metal net around the beach is set in motion. According to a captain of an ocean-going vessel interviewed as an authority in the field, the net is not necessary since it is easy to frighten any shark "by shouting as loud as possible, and by striking the water with one's feet and hands". Everything moving in the water is aimed at, with guns, pistols, spears and oars. Finally, in the midst of this hysterical war, the voice of academic reason makes itself heard when Dr Frederick Lucas, director of the Natural History Museum, declares: "No shark could skin a human leg like a carrot, for the jaws are not powerful enough to induce injuries like those described by Colonel Schauffler."

The experts having spoken and equivocation having cost 250,000 dollars in loss of revenue for the seaside resorts, there was still hope that this could be made up over the six remaining weeks of the summer, with the final blessing of the fisheries department in Washington. In fact the person in charge there declared that the two attacks were without any doubt attributable to the same shark, which would have been driven to attack Van Sant as it had lost its way far from any zone with plenty of fish; having tasted human flesh, it would have continued swimming near the coasts until satisfying its appetite with Bruder. Doubtless the situation would not recur.

Matawan, again in New Jersey, is a small inland port 16 kilometres from the Atlantic Ocean, to which it is connected by a narrow creek just a few metres wide at low tide. Its wharves, long since disused, serve as diving-boards every summer for the children of Matawan.

In early July 1916, Rennie Cartan, aged fourteen, dives into the muddy water of Wyckoff Dock. He is scarcely in the water before he feels a whiplash at stomach level, as if he were being violently rubbed with coarse sandpaper. Hastily climbing back on to the quay, he discovers a superficial bleeding wound and warns his companions: "Don't go in! There's a shark or something!". Nobody takes any notice of the warning and the incident is forgotten.

On 11th July, a few kilometres away, a fisherman catches a 3 metre shark, something that has never been seen in the region, but no more is spoken of it.

On the morning of 12th July, Captain Cottrell, retired seaman and casual fisherman, is walking along the new bridge which crosses Matawan Creek 1.6 kilometres downstream of Wyckoff Dock. Eleven days have passed since Charles Van Sant died 120 kilometres away, and six days since Charles Bruder was killed 40 kilometres from Matawan. Cottrell suddenly catches sight of a grey shadow gliding slowly upriver beneath the bridge, carried by the rising tide. He shouts in the direction of two workmen a little farther on, who also see the shadow passing. He then runs to telephone the barber at Matawan, who is also the chief of police, and hurries into the main street warning everybody to stop their children going to the creek where they swim every day. Everyone bursts out laughing at the idea that a shark could come prowling inland in a creek no more than 12 metres across at its widest point, and Chief Mulsonn does not even leave his barber's shop. Captain Cottrell therefore returns to the creek.

One of the stores that he has forewarned is that belonging to Stanley Fisher, a giant of twenty-four years who has just set up as a dry-cleaner at Matawan, even though he could have followed in his father's footsteps and joined the navy. Many regard it as lamentable that a man so big and strong should content himself with such an occupation, but hasn't he got the future ahead of him to change direction?

That 12th July is decidedly hot, and young Lester Stilwell is impatiently waiting to leave his father's mill where the heat is almost unbearable. His father releases him from work for the afternoon and he immediately goes and joins all his friends on the edges of the creek.

Later that afternoon his friend, eleven years old Albert O'Hara, is about to leave the water when Lester calls to him: "Look at me floating!". Albert turns to him in surprise. Lester is so thin that he usually has difficulty floating without splashing about. At that moment, something hard and rasping barges into Albert's right leg. He looks beneath the water and catches a glimpse of the sinuous tail of an enormous fish. His friend Van Burnt also sees it, the biggest, the blackest he has ever seen. They call Lester, who answers them with a yell. Van Burnt catches sight of the body of the enormous fish turning around as it seizes Lester; it is indeed black above, but it has a white belly and enormous pearly teeth. He is sure now that it is a shark that has just shut its jaws on Lester's frail body and dragged it beneath the reddening waters of Matawan Creek. Lester will never shout again. All the children dash out of the water, and some run to the village to give the alarm while the others despairingly call Lester.

There is nothing but panic and screaming on the banks where Captain Cottrell took his walk the day before.

Among the adults who are running towards the creek without knowing exactly what has happened is Stanley Fisher, who has taken the time to slip on his swimming costume.

The schoolmistress Anderson warns him: "Remember what Captain Cottrell said, it could be a shark!". Fisher stops for a moment. "A shark? Here?" He seems huge as he stands in front of the schoolmistress, thinking out aloud as if to convince himself. "Too bad, I'm going anyway." He immediately heads for the little creek, where two hundred people are now gathered, including Lester Stilwell's parents. He tells two men to get a boat and tow a stuffed chicken towards the other end of the creek, hoping in this way to lure the monster from the area he himself must search. Fisher knows that there is an underwater cavity in the creek, and he is sure that the shark is hiding there with Lester's body. Once his archaic plan of action is put in place, Fisher plunges towards that hole. When directly above it, he takes two big breaths and then disappears underwater.

The detective Van Buskirk arrives just in time to see Fisher dive. The surface remains calm for about twenty seconds before a big swirl seems to announce Fisher's return to the open air. But instead, the surface becomes calm again and clouds over with a rapidly expanding red stain. Van Buskirk hurries by boat towards the sinister stain, from which Fisher's head, then his chest, slowly emerge. Seen from a distance, Fisher seems to be standing beside the hole, with water reaching to his waist. He turns his back to the crowd, who therefore do not see the spectacle that greets Van Buskirk when he reaches Stanley Fisher. The latter is staggering, holding in both hands the bleeding remnants of one of his legs. Van Buskirk barely has time to grab him by the shoulders before he slumps face downwards. He can only hoist him up by the waist while the helmsman makes a half-turn towards the dock. The crowd then get a view of Fisher as a macabre ship's figurehead. His body is out of the water sufficiently for the hideous wound to be visible. From the hip to the knee, all the flesh has gone from his right leg, which is now only joined to the trunk by the femur, itself deeply gashed along its whole length. Reaching land, Van Buskirk manages with difficulty to stop the flow of blood escaping from the torn femoral artery using a piece of rope. Fisher makes desperate efforts not to sink into unconsciousness, as if he definitely wants to say something. He is taken on a makeshift stretcher to the station, where there is a three-hour wait for the next train. There a doctor manages to ensure that the bleeding from the wounds has been arrested, but a further three hours' travelling is necessary before finally reaching the hospital. Despite his pain Fisher remains conscious until reaching the operating table, where he succeeds in delivering his message. He did find young Lester's body at the spot he had envisaged, and he did manage to snatch it from the jaws of the shark before being attacked himself. Fisher succumbs even before being given the anaesthetic.

While Fisher was waiting on the station platform, several boys continued to bathe a kilometre downriver from Matawan Creek, unaware of the drama that had just taken place. When they were at last informed, they all rushed out of the water. Joseph Dunn, the youngest of them, was the last to use the dock ladder at Key Port. As he was starting to climb up, he felt something like a huge pair of scissors lacerate his right leg: "I felt my leg inside the shark's mouth, I thought it was going to swallow me whole." His brother Michael and two other older boys clung on to him in a deadly tug-of-war against the shark, which refused to release its hold. They tore open Joseph Dunn's flesh but saved his life. The shark finally released its victim, the third in less than an hour, and the only one to escape with his life even though he had to have his leg amputated at the thigh.

The tragedy was followed by one of the most intensive shark hunts ever seen. Hundreds of volunteers flocked in. After the creek had been shut off with metal nets, hundreds of kilos of dynamite were set off everywhere where the shark could have hidden itself, but nothing significant was caught. Even small craft fitted with cannons for harpooning whales were brought in. Several sharks were caught here and there, immediately stuffed and put on show for the public, in return for their participation. Meanwhile, Lester's body was recovered a hundred metres from the spot where he had disappeared, bearing seven wounds including two to the abdomen.

Two days after these events, Michael Schleisser, a taxidermist, captured a shark 2.6 metres long off South Amboy, six kilometres north of Raritan Bay. When he opened it up, he found seven kilos of flesh and bones, very quickly identified as of human origin, in its stomach. Among the remains was part of a bone apparently belonging to Charles Bruder attacked nine days earlier. Schleisser mounted the shark skin so as to exhibit it and it was definitively identified as a Great White Shark, *Carcharodon carcharias*. Following this capture, the attacks immediately ceased, confirming that this shark was a loner, the one and only perpetrator of the five attacks.

There was no lack of theories to explain this phenomenon. Some people claimed that it was the time of the year for sharks. Others suggested that the beast must have been suffering from a kind of mange in the way dogs do, when they can be driven mad by it. It was also thought that because of the war sharks were no longer finding the usual food that was thrown overboard from liners, and were falling back on other sources. Recent maritime disasters had also perhaps given these unscrupulous predators a taste for human flesh.

Whatever the real reason may have been, these events demonstrated for the first time the destructive power of a shark capable of attacking man.

TERMINOLOGIE ANATOMIQUE GÉNÉRALE DU REQUIN.

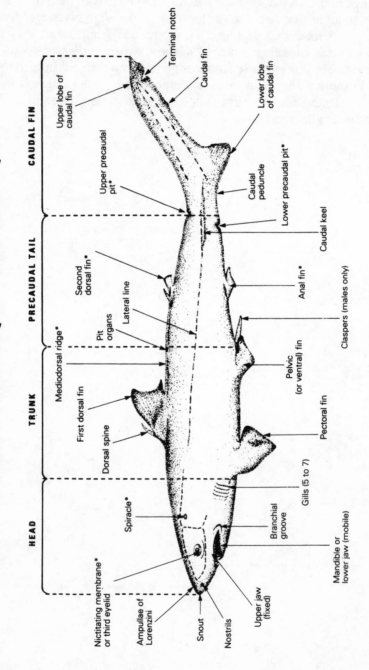

HEAD   TRUNK   PRECAUDAL TAIL   CAUDAL FIN

Terminal notch

Caudal fin

Lower lobe of caudal fin

Upper lobe of caudal fin

Upper precaudal pit*

Lower precaudal pit*

Caudal peduncle

Caudal keel

Second dorsal fin*

Anal fin*

Lateral line

Claspers (males only)

Mediodorsal ridge*

Pit organs

Pelvic (or ventral) fin

First dorsal fin

Pectoral fin

Dorsal spine

Gills (5 to 7)

Spiracle*

Branchial groove

Nictitating membrane or third eyelid*

Mandible or lower jaw (mobile)

Ampullae of Lorenzini

Snout

Nostrils

Upper jaw (fixed)

* Organs not found in all sharks

44

3

AN EXTRAORDINARY MACHINE

DEVILISH JAWS

A FLAWLESS SET OF TEETH

THE SHARK'S JAWS exhibit characteristics unique in the animal world, no terrestrial predator has jaws that come anywhere near matching them in their perfection.

The ancestors of modern sharks were equipped with an upper jaw fixed to the skull (as in man) and with a mouth located at the extremity of the head. Since the age of these fossil sharks, which had to content themselves with relatively small prey, the mouth has moved underneath the head, and the upper jaw, disengaged from the skull, has become mobile. The shark is therefore equipped with two mobile and independent jaws, enabling it to swallow much larger prey and to tear off enormous pieces of flesh.

The spectacular distance these jaws can open would be nothing were they not subtended by muscles of exceptional power. The shutting force of a shark's jaws was measured a few years ago using an apparatus tested by J.N. Snodgrass which he called the "gnathodynamometer". With the help of Dr Perry Gilbert he measured the dentitional strength of the Tiger Shark, the Lemon Shark and the Dusky Shark. The maximum force recorded for a single tooth of a Dusky Shark was 600 kilos per 2 mm^2, or 3 tonnes per cm^2. Even then it should be noted that their specimens were not

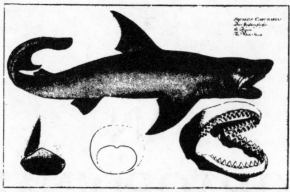

The Great White Shark as seen by Bloch in 1785 in his *Histoire naturelle du poisson*. This encyclopedia entry served as a model for a very long time and even its errors were taken up (in particular here the inversion of the upper and lower teeth).

more than 3 metres long, so doubtless this enormous force must be doubled for animals of 5 or 6 metres in length. As an indication, the strength exerted by a human jaw is 220 kilos per cm^2 for a man weighing 70 kilos (11 stone). It must also be remembered that, in addition to cutting, these jaws are capable of crushing bones.

Besides its extensibility and its Herculean strength, the jaw of the shark possesses a third formidable feature: a set of teeth whose anatomy and manner of replacement are also unique in the animal kingdom. The shark's teeth can be renewed indefinitely as fast as they are lost either at the time of an attack or when they fall out spontaneously. Several sets of reserve teeth exist behind the set that is functional at the time, i.e. the one that sits on the ridge of the jawbone. There are thus at least five sets, covered to varying degrees with buccal mucous membrane depending on whether they are number five, four, three or two in line. If one or more teeth in the functional row are exposed at the root, broken or torn out, the corresponding tooth or teeth of the next row will move up and become functional.

Furthermore, the teeth are held fast on a very strong fibrous tissue, allowing them to be erected when the mouth is opened. The opening movement makes them turn forwards and outwards, enabling the shark to bite firmly and to hold on to what it bites. Taking into account the fact that each one of these teeth itself bears smaller teeth, making it a veritable saw, and that its concavity at the rear gives it the qualities of a hook, one can understand why the wounds left by such a formidable jaw are always dramatic, if not fatal.

Even when dead, the animal remains a potential danger to guard against. A jaw that has been cut away from its surrounding muscles, and placed on the deck of a boat to dry in the sun has a good chance of suddenly snapping shut again owing to retraction of the fibrous articulations. Woe betide the hand or the arm that lingers too long; accidents have occurred on a number of occasions.

For the same reasons, the best way of removing the hook from the shark's mouth is to wait until the following day. Apparent death in this animal is often only an illusion, even several hours after it has been out of the water.

Our information on the rate of tooth replacement in the sharks stems in particular from the studies done by the American Samford Moss on species of scavenging shark. The Lemon Sharks, for example, renew their teeth every 8.2 days for the lower jaw and every 7.8 days for the upper jaw. This is doubtless an extreme case, and it seems that the largest species have a longer cycle, of the order of six months to a year. It has been possible to demonstrate this in a crude fashion through teeth gathered on the bottom of certain large aquaria.

The shape of the teeth is highly characteristic according to the order to which the animal belongs, and thus not only allows for the easy identification of a shark once captured, but also aids the identification of the species of shark responsible for an attack, when all or part of a tooth is found in the victim's wounds.

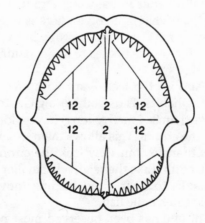

DENTAL FORMULA
This helps in the identification of a shark. 13/13, for example, means 13 teeth in the upper half-jaw and the same number in the lower half-jaw. In certain species (particularly in the requiem sharks), a few very small teeth can be seen in the centre of each jaw, causing the formula to vary.

SEQUENCES OF A BITE

At the time of an attack, all these anatomical structures are extremely mobile and the dynamics of the bite are very particular for a classic surface attack (as the illustration shows).

Feeding on large fish or marine mammals involves tearing away whole mouthfuls. Contrary to common belief, the shark does not roll onto its back or side to attack, but approaches and attacks from an oblique angle. First it plants the teeth of its lower jaw into the flesh of the prey, and then, in a second movement, the teeth of the upper jaw. This sequence is perfectly logical when we look at the shape of the teeth in the jaws (see the directory at the end of the book). The most pointed ones are always in the lower jaw, which is therefore the most suited for piercing the flesh and maintaining a grip while waiting for the blade-like cutting teeth of the upper jaw to come slicing into the prey, which can no longer escape. Certain sharks such as the Sand Tiger Shark (known as the Grey Nurse Shark in Australia and the Spotted Ragged-tooth Shark in South Africa) are equipped with very long and pointed lower teeth curving backwards like hook-shaped fangs. Once the teeth have sunk into the prey, the shark shakes its head sideways violently, causing a real "sawing" of the flesh. In this way a big shark can remove about 20 kilos from a large marine mammal in one mouthful.

Coppleson has observed that certain sharks "relax" their body after biting, so that it hangs in the water and its weight helps to tear the flesh. I think that what is involved here is a fairly static posture for which we cannot make a standard rule, but which could be that of scavengers feeding on the bodies of large mammals floating at the surface.

Among some species the teeth vary with age, leading to certain preferential feeding habits. Thus both young Great White Sharks and the subadults (less than 3–4 metres long) have a longer and narrower tooth shape than the adults. This allows them to seize small bony fish and other elasmobranchs, but they lack the cutting capacity of the broad teeth of the adult. Like the relatively smaller Makos, the young Great Whites, thanks to their rapid turning movements, are more able than adults to catch agile fish.

Another type of bite has been observed most notably from Great Whites on large pinnipeds, such as elephant seals or sea lions, but also on other sharks and even human victims. This bite is not made in order to remove a piece of flesh, but to intimidate, or to taste, or even to ward off the counter-attack that a big pinniped is capable of inflicting with its teeth and its claws (some can attain 5 metres and

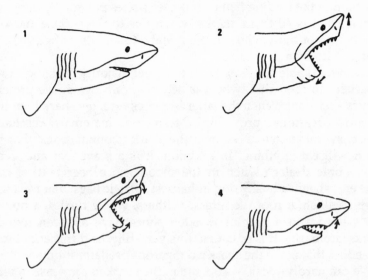

1. Just before attacking, the snout and the lower jaw are in their normal resting position and the teeth are more or less parallel to the floor of the mouth. If the prey is small, the teeth do not need to be brought into action, and it can be swallowed immediately without the following sequence being necessary.

2. The snout is raised, and the depressing of the lower jaw permits a maximum opening of the whole mouth.

3. Without the snout changing position, the upper jaw itself turns forwards and downwards, exposing the upper teeth; the backward-angled direction of these teeth helps to get the prey into the mouth.

4. The snout is lowered, after the prey has been seized, allowing the upper jaw to return to its normal position beneath the skull (small arrow).

All these sequences follow one another at an extremely rapid speed, in 1/400 of a second, this has been captured by cameras with ultrarapid action.

weigh up to 2700 kilos). In all cases this bite is made with the upper jaw only, as if the shark had decided *a priori* to bite, but not to tear. Numerous stories reported in this book testify to such a mode of attack against man, which should perhaps lead us to label the shark as a "man-biter" rather than a "man-eater".

FEEDING, APPETITE AND DIGESTION

The shark's usual nutrition is discussed throughout the course of this book, in particular the feeding methods of certain benthic (bottom-dwelling) species and of the three sharks equipped with filters like whales. In this chapter I shall confine myself to mentioning the feeding habits of those species potentially dangerous to man, that is all those sharks which rise to the surface and come near coasts, or those whose rapid metabolic activity brings them into contact with man.

The remarkable adaptability of the great majority of sharks applies especially to their diet, which is selective only when the preferred prey is abundant. When the latter becomes rare, the shark then turns towards occasional prey: birds, terrestrial mammals, discharged refuse, swimmers etc. Like man the shark is omnivorous: that is to say it will eat anything. In addition, it is a scavenger and even a "bric-a-brac dealer" when in the mood, since beer bottles, coats, chickens, sheep, a husky dog in harness, plastic bags, a horse's head, large shellfish, a roof tile, grass, feathers, whole turtles, a hornless cow's head and a pair of wooden shoes have all been found in sharks' stomachs. It is of course this versatility and this adaptability in feeding that make the shark a dangerous predator for man.

We can hardly speak of appetite in the shark in the sense in which we understand it for man, with precise timetables, a constrained periodicity, a necessary variety, and a feeling of satiety at the end of taking in food. Here again, it adapts according to the resources of the environment.

If there is one prey the Blue Shark is especially fond of, it is the squid. It so happens that squid regroup in the spawning season on moonless winter nights, at certain favoured places. These gatherings involve millions of individuals coming together, for instance to the west of the island of Catalina 40 kilometres from Los Angeles. After having been fertilised, the females deposit their eggs on the sandy bottom and shortly afterwards die, as do the males. This is why the beds are sometimes covered with several metres of squid bodies, piled one on top of the other. Among the many carrion-eaters and predators that come to feast every year on this veritable godsend are the Blue Sharks, which simply swim into the banks of squid with mouths wide open. They stop only when their engorged stomachs can hold no more, and the squid dangle from each side of their mouths. The shark often even regurgitates part of the food, only to start eating the same huge quantities again. This voracity allows them to wait for the weather conditions necessary for another opportunity, which can be very late in coming. Some species have thus been seen to fast for months, and it is thought that they survive thanks to the oil reserves contained in their liver. This last hypothesis seems to me to hold good if we note that sharks' livers can reach 25% of their body weight (see "Hydrodynamics") and contain 90% oil. This represents, on the basis of a calorific value of 9 calories per gram, reserves of around 360,000 calories for a 500 kilo shark. To give some idea of what that can represent, the basal metabolism of a sedentary man weighing 70 kilos (11 stone) is of the order of 2,000 calories per day.

In captivity, with all that that implies in the way of differences compared with the natural environment, sharks have been observed to consume a daily ration equivalent to 0.4% to 2% of their own weight.

Great White Sharks are equally capable of ingesting enormous quantities of food, if ever the opportunity presents itself. Thus it has been calculated that the great mass of blubber they ingest when they swallow the corpse of a whale in a single meal, covers their calorific requirements for two months. Up until a few years ago, Durban was a very important whaling port. The whales were harpooned and killed about 200 kilometres offshore, before being filled with compressed air beneath the skin so as to be towed towards Durban. Between the moment when the whales were harpooned and that at which they were towed, a great many sharks inflicted considerable damage on the carcasses, ripping away enormous pieces of blubber and often reaching the flesh. At times the attacks were so vigorous that the carcasses were perforated, allowing the compressed air to escape, and resulting in the animal being lost. The financial losses were equally heavy, but no solution was found before the cessation of the whaling industry.

Another example of the feeding adaptation of sharks in a regional habitat is that of the Great Whites which are found along the Californian coasts and whose numbers increase as the colonies of elephant seals there get bigger. It is known that Great Whites prefer pinnipeds to any other prey, which explains their abundant presence in south Australia and north California. In this latter region, where they were studied in detail by John Cosker (1986), an increase in their numbers has run parallel with an increase in the number of attacks on surfers, whose silhouette at the surface resembles that of a pinniped.

It has long been asserted that hunger is the principal force behind unprovoked assaults on man. If such were the case, we could say, like Balbridge and Williams, that the shark or sharks implicated are certainly poor hunters. We need only look at the number of survivors of such attacks (more than 50%) to see clearly that hunger must only exceptionally be the reason behind the attack. What after all does the frail carcass of a man count for against the veritable mincing machine with which the smallest shark is equipped? Why would a predator so formidably armed change its mind in the process of biting, suddenly deciding not to eat its prey? Why, conversely, are certain rare attacks conducted with a savagery that gives the victim no chance, as if the shark had indeed decided to eat?

It is certain that hunger motivates the attack in a minority of cases, but does that mean that the shark is never hungry and never swallows anything? Studies have been made at natural aquaria by Graeber (1974) and Longval (1982) which showed a certain periodicity in feeding by the Lemon Shark, with some peaks every four days and others every twenty-eight days (lunar periodicity?). They showed that, once the animal has eaten its fill, it needs four days for it to be hungry again, and above all that going without food and satiation play an important role in its feeding behaviour.

It can be said that sharks know hunger, but that hunger is only exceptionally the reason for their attacks on man. The majority of sharks have an activity rhythm that is nycthemeral (covering one day and one night), but also seasonal. A corresponding dominant diurnal activity for certain species accounts for greater numbers of shark-human interactions involving these species.

The shark's digestion is another very particular biological process, oriented towards a vigorous predatory capability. Whereas the digestive tract of a man 1.8 metres (5 feet 10 inches) tall reaches 9 metres in length, that of a 3 metre shark does not exceed 2.7 metres. When the food has been subjected to the action of the gastric juices with their relatively concentrated hydrochloric acid base, it passes into the intestine, which comprises a spiral-shaped valvule, in other words an arrangement that permits a maximum absorption surface within a minimum length (always with the objective of preserving maximum space for the liver, the stomach and the embryos).

Studies carried out on several species have shown that, if the initial phase of digestion was relatively rapid (of the order of twenty-four hours), three to four days were necessary for complete digestion, and more in low ambient temperatures.

At 10°C the Spurdog (or Spiked or Spiny Dogfish) requires five days to digest herrings fully. The Porbeagle has a higher temperature, which enables it to have more efficient muscles and a more rapid digestion owing to a heat exchange in the region of the intestinal vessels.

There is an amazing characteristic peculiar to the shark, which was hinted at at the time of the "shark arm" affair, reported elsewhere in this book (see Chapter 7). This involves a delayed digestion which has been observed several times in the stomach area, allowing organic remains to be recovered intact after a week in the stomach of a shark, or half-digested after several weeks.

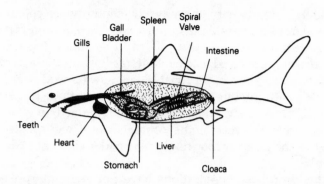

The digestive system of the shark

In September 1935, Patrick Quinn went overboard from a sugar-carrying ship at dead of night near the Queensland coast. Buoys were thrown to him, and he managed to hold on to one. The ship's lifeboat was launched, but, at the very moment it reached Quinn, he was suddenly snatched from his buoy, vanishing for good.

Three weeks later, on 31st October, the crew of the *Aldinga* captured a 3.5 metre shark. In its stomach several human bones were found, notably those of a complete leg still wearing its boot. A short time after, the crew of the freighter *Alynbank* brought up another shark which also contained undigested human bones. In both cases the remnants were identified as being those of the unfortunate Quinn.

On 5th January 1949, three amateur fishermen landed seven sharks within a few hours from their boat 6 kilometres off Leighton in Australia. These included a 2.5 metre Tiger Shark, in the stomach of which they found a small human hand, partly decomposed, mixed with some wallaby bones. After a few days, the police identified the remains as being those of the young Arthur Straman, aged seventeen, who disappeared on 27th December while bathing on the island of Lancelin. In this case, again, the director of fisheries in the region was extremely surprised at the incredible state of preservation of the hand after ten days. In fact it seems that in fact this constitutes the rule and not the exception, as a final, less morbid example demonstrates.

On 23rd August 1950, a Great White Shark was placed in the pool at Sydney Zoo, uninjured and in perfect condition. After a few days, it regurgitated several pine needles, a few tin cans and other refuse, following which it accepted the horse meat that was offered to it. This meat apparently did not agree with it, as it regurgitated it a few days later. After twenty-one days in captivity its colour became paler, a sign of a probable fatal outcome. The shark in fact died on the twenty-third day following its capture. From its stomach two dolphins were recovered, each 1.2 metres long and in a perfect state of preservation. They had probably been swallowed shortly before the shark's capture, but none can explain how these dolphins were preserved as if "new", nor how they remained in place in the stomach while the rest of the contents were regurgitated. In his report, the zoo director mockingly returned a verdict of "death from dolphinitis".

To this day these observations have not been interpreted, but I think that the partial regurgitations can be explained by peristaltic motions the force of which does not allow the expulsion of over-large prey. As for the state of preservation, this implies on the one hand an "intermittent secretion" of the digestive juices (by means of a reflex circuit triggered by endogenous and or exogenous stimuli) and, on the other, a spontaneous non-putrefaction resulting from a special substance secreted by the mucous membrane of the stomach. As these animals are cold-blooded and live in moderate temperatures, decomposition is certainly not accelerated by the internal temperature, but that is not enough to explain three weeks' preservation. I think, therefore, that we may imagine the secretion of a particular substance which has the properties of formalin, and with which the prey would be impregnated outside the periods of digestion (during which the hydrochloric secretion would dominate).

An extremely aggressive behaviour exists in sharks, which is known as the "feeding frenzy". This mode of feeding, which fortunately is not systematic, requires a number of factors to come together:

- the presence of several sharks of which at least one is hungry enough to bite

- the presence of a quantity of prey sufficient to warrant competitive behaviour

- a triggering stimulus.

This frenzy is not the appanage of the large specimens, which are not nimble enough, and which in general live solitarily. It is most often the act of small or medium-sized sharks belonging to species which are not necessarily among the most dangerous.

During the course of the feeding frenzy, any inert or living object is torn apart and swallowed indiscriminately, and the sharks may even devour each other if one of them is injured. All it needs is one of the sharks to start eating for the death-knell to begin. Even those which are not hungry will rush at the prey, substituting aggressiveness of competition for appetite.

A German man weighing 110 kilos (over 17 stone) was the victim of one such frenzy a few years ago in the Republic of South Africa, when he was attacked by sharks of less than 2 metres in length. When his remains were found a few hours later, they weighed no more than about fifty kilos.

EXCEPTIONAL SENSORY ORGANS

The formidable war machine that is the shark possesses organs for detection, for searching, for stalking and for identification which are at once numerous (eight), highly efficient (even at dead of night in murky waters), and in some cases unique in the animal world (the lateral line and ampullae of Lorenzini). The remarkable way in which they complement each other permits the approach of a shark to be compared with that of a modern torpedo, whose various electronic means of reaching the target make "contact" almost inescapable. I shall describe the various sense organs of the shark in the order corresponding to their respective linear ranges of sensitivity.

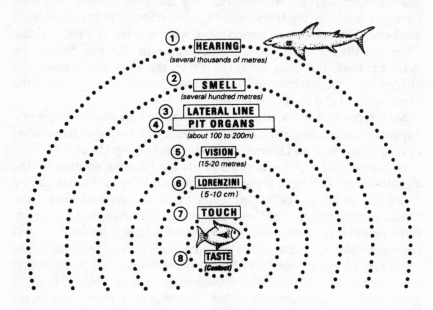

The sensory organs and their respective ranges of sensitivity

HEARING

The hearing spectrum of sharks extends from 10 Hertz in the low frequencies to around 1000 Hertz at most in the high frequencies. Considering that that of man ranges from 25 to 16000 Hertz, it is obvious that a lot of high-pitched sounds can be heard by man that are not heard by the shark. On the other hand, the small difference between 25 and 10 in the low-frequency register indicates that the shark can discern many more bass tones than can man, and this difference in the sphere of low-pitched sounds largely compensates for the difference in high-pitched ones, as the figure demonstrates.

Experiments conducted by Myrberg, Nelson and Johnson in the last twenty years have shown that sharks can detect very low frequencies inaudible to man, and that in addition the sources of these frequencies attract them. Microphones were positioned on sea beds emitting artificial sounds or recorded natural sounds – these were of several origins: the sounds of fishes in normal situations and in dense groups, the sounds of fishes having taken the bait and having been harpooned, and diverse musical sounds.

All these noises attracted the majority of the sharks in the surrounding area. They arrived in the immediate vicinity of the source in the space of 30 seconds to 1 minute, coming from very much farther than the limit of visibility (20 to 30 metres), and thus attracted solely by the transmitted sounds (inaudible to the research workers). The subjects of experiment were species as varied as the Thresher, the Silvertip, the Oceanic White-tip, the Bull Shark, the Black-tip Reef, the Grey Reef, the Tiger, the Blue, the Lemon, the Sharpnose, the White-tip Reef, the Mako, the Nurse, the Hammerhead and the Bonnethead.

Such rapid and systematic reactions explain why sharks frequently appear, coming from any direction, after a fish has been harpooned or caught on a line and immediately starts splashing about.

Non-synthetic low-frequency sounds are likewise emitted by the vibrations of fishing cables and lines and by a human being splashing about in the water. Thus these actions will attract the sharks in the vicinity. Another parameter of attraction to a frequency is the sound level. Frequencies transmitted at high intensity attracted the same sharks from much farther away. Once in immediate proximity of the source of emission, some sharks fled if the sound level was suddenly increased (20 decibels or more, which is not a great deal in absolute value but constitutes a fairly significant increase gradient).

Comparative scales of the hearing ranges of man and of the shark with regard to the piano. The shark's register is much wider in the field of low frequencies.

It is already possible, from these observations, to derive preventive measures to avoid the curiosity of sharks: do not make any noise underwater (even any of those we cannot hear) and kill captured fish swiftly so that they no longer struggle. The current of water around the taut cable of an anchor certainly forms a sound source liable to attract a nearby shark to a pleasure boat, and it seems good advice to me to be on the lookout for sources of low-frequency emission of this kind before immediately diving from the deck of a boat in risky tropical waters. Several times, when diving from a yacht which had only just anchored, I have come face to face with sharks which showed no concern as if, identification having been made, the boat was no longer of interest to them.

A source of low frequency whose sound level could be suddenly increased ought then to be investigated as a possible means of repelling sharks. Although the big ones are less influenced by this than the small ones and often keep on coming, incapable of stopping, they are nevertheless liable to change direction at the last moment.

This twin power of one and the same deep sound to attract and then, when turned up louder, to repel is the reason for contradictory interpretations in the specialist literature.

If we take into account the reality of these acoustic stimuli and the excellent propagation of sounds in water (1500 metres per second), hearing can be considered to be the first sensory function of the shark to be alerted by a distant object, and the range at which it is alerted is the greatest, of the order of several thousand metres.

SMELL

The shark was at one time called the "nose of the sea", and it is true that smell ranks second in sensitivity among the shark's various sensory organs, with a detection range of about 500 metres. The nostrils do not open into the mouth as in the frogs but are located behind two folds of skin known as the flaps of Schneider, so that swimming creates a constant flow of water over the sensory cells inside the nasal cavity.

Unlike man, the shark possesses nostrils that are different from each other, and it heads in the direction of the side that receives the strongest sensory impulse. This directionality is augmented in the course of swimming by the horizontal balancing movement of the head, which allows the nostrils to test a wider corridor of smell.

The sensitivity of this organ is spectacular, since sharks are capable of detecting 1 part of mammal blood in 100 million parts of water. A few years ago, at the Lerner Marine Laboratory at Bimini in the Bahamas, the sense of smell of the Lemon Shark was tested. The experiments demonstrated that these sharks are sensitive to extracts of tunny in dilutions of 1 part in 400 million. The American, Albert Tester, conducted experiments in the Marshall Islands in the Pacific under conditions of progressive fasting. He established that, deprived of food, the animals respond positively to concentrations of 1 part in 10,000 million. He also noticed that these sharks have a profound aversion to human sweat, which could be a factor explaining certain apparently selective attacks on some bathers rather than on others.

So, when swimming, the shark continuously receives a host of odours, some of which trigger off the hunting procedure. Apart from the odours already mentioned, there is a more subtle one that also provokes pursuit, that of the secretions released by frightened fish.

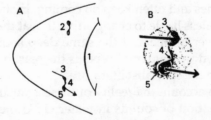

Characteristic arrangement of the nostrils in the shark allowing detection at great distances:

1. Mouth 2. Nostrils 3. Opening for inflowing current 4. Wings of nose 5. Opening for outflowing current. The shark's forward movement causes water to penetrate through the entry orifice of the nostril, it then passes into the nasal capsule and leaves again by the exit orifice (A). Inside the nasal capsule are numerous folds which increase appreciably the number of olfactory receptors exposed to odours carried by the water, thus enhancing the sensitivity (B).

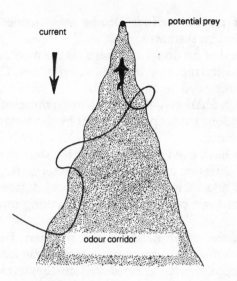

How a shark detects a prey by its odour

This was demonstrated in 1960, again by Albert Tester, in Hawaii. He placed whole fish of the same species in two container tanks A and B. In tank B, he panicked the fish by shouting and beating the walls with a stick. He siphoned the water from container A into a shark pool and the sharks showed very little reaction; but the water from container B, which had held the panicked fish, instantaneously provoked intense hunting activity, some sharks even going so far as to bite the outlet tube of the siphon.

So what we see from these experiments is that the shark will be fiercely attracted by a multitude of odours for which its hunting instinct has been genetically programmed; but also by others with which it is not familiar but which its curiosity incites it to identify at closer quarters, such as a dinghy with the blood or vomit of shipwreck victims leaking from it. In such circumstances a point should be made not to throw anything overboard that is not carefully sealed in plastic bags. We shall see that one of the methods of avoiding the attention of sharks is the "Johnson bag", which traps inside all the odours which could emanate from the shipwrecked person who has fallen into the sea.

This chemoreception is not only effective with regard to electrolytes, amino acids and amines, but equally allows sharks to distinguish clearly waters of different salinities (Hodgson, 1978). This capacity could explain the different displacements and the geographical segregations that are observed within the same populations at certain periods of the year (zones reserved for gestating females, for instance).

It is olfaction that no doubt also permits the mechanism of sexual segregation, so often reported in studies of fisheries. This mechanism has been demonstrated again recently from direct observation of male Grey Reef Sharks (*Carcharhinus amblyrhynchos*) which were moving along odour trails apparently left by the females of the same species (MacKibben).

Some people have tried to derive from the statistics on attacks on human prey of different races, data on whether or not the frequency or otherwise of attacks is affected by race and different odours. But so far it has not been possible to prove anything important in this field.

It seems, on the other hand, that more men than women are attacked. It is evident that many more men come into contact with sharks, but, if the figures are reduced to the number of attacks relating to the same number of men and of women in the water under the same conditions, it seems in actual fact that women are less likely to be attacked. Should we be talking about different odours (apart, of course, from the menstrual periods during which women must never bathe in waters where sharks may be present), then perhaps the olfactory organ of the shark is the most selective in the animal world. The question would merit scientific exploration.

THE LATERAL LINE

Just as we were able to talk of chemoreception in relation to sense of smell, we now come to mechanoreception, the amplification by specialised organs of variations in pressure. These variations in pressure can be the vibrations from a frightened fish splashing about, from a ship's propeller or from the flippers of a swimmer at the surface. The lateral line acts in the manner of a movement detector by picking up the smallest movement in the water and the vibration resulting from it.

This system seems to take the place of the ear for perception of low frequencies and in addition plays a determinant role in the fish's balance. As long as fifty years ago, an American called Parker, of Harvard University, without knowing the histology of the system, had shown that the Spotted Dogfish (a small shark also found on European coasts), deprived of its auditory and visual organs, could still perceive disturbances in the water since its lateral system remained intact. On the other hand, as soon as he cut the nerves of the lateral system, the dogfish showed no reaction.

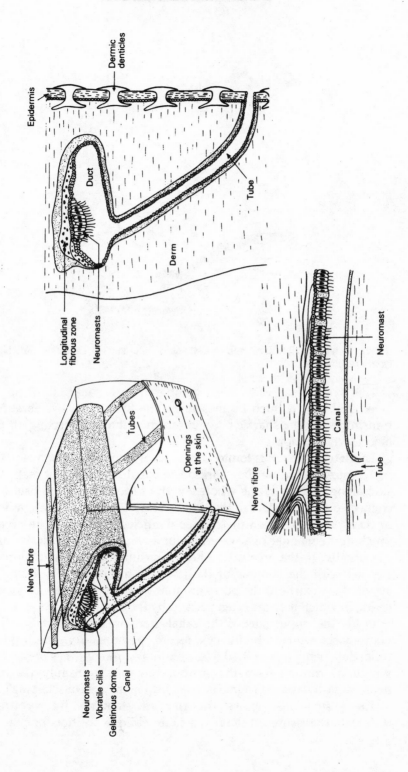

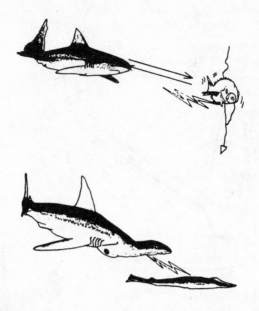

(ABOVE) Detecting vibrations from a distressed fish. (BELOW) Detecting heartbeats from a resting fish.

This system has often been compared to a sonar, allowing an object to be pinpointed by measuring the time taken by vibrations transmitted by the shark's body to return to it after reflecting off the object (echolocation).

This lateral system is found only in a very small number of aquatic vertebrates, including the sharks. The canal itself runs beneath the skin along practically the whole length of each side of the fish. It is connected to the surface of the skin by small tubes and the alignment of their pores (between the dermic denticles) makes up the organ, much in the manner of a waterline for a ship. Other identical canals are situated in the area of the head, skirting the eyes and running parallel with the contour of the jaws. In the bony fish, a similar system may exist but the pores are more visible than they are in the shark, in which they are often masked by the denticles of the skin.

Inside the upper part of the canal there are sensory cells called neuromasts, whose vibratile cilia record pressures from the ambient fluid, data being transmitted to the brain by afferent nerve fibres. The vibrations coming from the surroundings are transmitted to the neuromasts by means of a mucus contained in the system's "piping".

The system also plays the role of stabiliser by recording differential pressures: if the fish is in an oblique position, one of the two lateral lines will record a higher (or lower) pressure on one side than on the other, and, by reflex circuit, the cerebral structures will order the actions necessary to the fins to re-establish identical pressure on each side and thus to re-establish the fish's balance. The role of this organ in the co-ordination of swimming, however, is not yet fully clarified. In 1977, Campbell confirmed that fish could still swim even if the nerves reaching the lateral system were cut, no doubt because vision and hearing compensate for this deficiency. This confirms the complementary and even the supplementary nature of the shark's sensory organs, for ensuring perfect swimming as well as for detecting hidden prey.

This system also enables a shark to detect a fish at rest, by its heart beats reverberating in the water.

THE PIT ORGANS: "TASTE BUDS" OF THE BACK

The shark has yet another extraordinary sensory organ at its disposal: the pit organs, or free neuromasts. The pits are used for discerning variations in the chemical composition and salinity of the surrounding water. They are closely linked to the denticles of the skin and are distributed all along the body, from the head to the caudal peduncle.

For each pit, two adjacent scales (or denticles), different from the others, overlap in such a way that between them they create a small pit or cavity. In young sharks, these differing denticles form a clearly visible ridge, while in the adult they are buried among the others. These pits are more numerous in open sea or pelagic species.

Inside each pit are a papilla and a large sensitive cell. What is remarkable about the device is that it is totally analogous with the papillae and cells which make up the taste organs on the tongue of man and other animals.

Even if analogy of function does not necessarily follow analogy of structure, the similarity in the present case is so absolute that we may with some justification make presumptions on the very probable role of these organs. Some people will argue that to have taste buds on the back or the flanks is very odd, but nevertheless experiments have confirmed that these structures are sensitive to chemical stimuli and this can only reinforce the idea that "organs of taste" are indeed present here. This also distinguishes this structure from the two other mechanoreceptors: the lateral line which we have just mentioned, and the ampullae of Lorenzini.

We shall see in the lead-up to attacks the deciding importance of variations in salinity, due for example to major waterfalls. In such cases the lowering of salinity in the coastal waters attracts sharks, whose "genetic memory" has recorded that this signifies a surge of organic debris from rivers and streams. In the course of evolution, the shark's genetic heritage has been enriched by the development of this detector, which has opened up new hunting grounds for it. The detecting range of the pit organs is comparable to that of the lateral line, of the order of several hundred metres.

AN EXTRAORDINARY EYESIGHT

As visibility underwater is limited on average to about twenty metres, it has long been considered that the shark must have pretty poor vision, no doubt adapted to its environment in accordance with the law of evolution which states that natural selection retains only the genetic mutations that are useful for the survival of the species. To read the specialists of the 1970s, compared with man, the shark was an animal with a real "visual handicap". It is hard to imagine, however, why man should have a better underwater vision than the shark: to each his own environment.

Doubtless for the same reasons, it was peremptorily asserted that the shark could not distinguish colours, for, underwater, beyond about ten metres, red disappears and the depths become uniformly greenish-grey and almost dark. The only accurate observation of the early specialists was that sharks can distinguish contrasts very well,

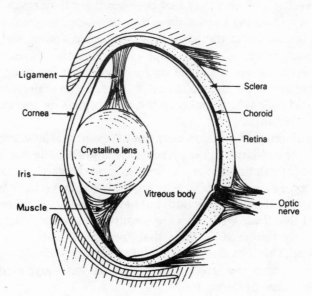

The shark's eye is very close to that of man

but even then it was believed that the shark cannot see anything at night. These preconceived ideas were all the more difficult to get rid of in some countries because they were relayed by certain leading figures, including one particularly popular in France.

We shall see that the shark's ability is as outstanding in the sphere of vision as it is in other areas, with or without sunlight, with or without colours, by night as by day, close range or long distance, under the water as above it.

As long ago as 1960, Perry W. Gilbert, at Bimini in the Bahamas, placed covers over the eyes of several Lemon Sharks in order to render them temporarily blind. When subsequently returned to the tank of the marine laboratory to compete for food alongside sharks with normal vision, they appeared incapable of locating a visually detectable prey. At the same time, in the same laboratory, he showed that sharks are attracted by bright shiny objects whereas they do not notice dull-coloured objects. Previously, Alain Bombard had noticed during a sea crossing that he could not put his yellow oilskins on the decks of his ship *L'Hérétique* to dry, as they were regularly snatched by sharks and other big fish.

A number of similar observations confirm this attraction of sharks to yellow or orange, and it is quite incongruous to see life-rafts still manufactured in which the part that gets immersed is yellow or orange, as if some manufacturers believed that their clients would only be shipwrecked off the Isles of Scilly rather than in the tropics.

The shark's eye is quite close to the standard eye of vertebrates, and the pupil can dilate and contract very rapidly, contrary to another popular belief. Gilbert has shown for the Nurse Shark that the maximum dilation of the iris, when exposed to darkness, occurs within 24 to 30 seconds, while maximum contraction during exposure to the light occurs within 5 to 13 seconds.

This means that a shark charging at prey which it catches sight of 20 metres away at the surface will, even though emerging from the depths, have time to adjust to the light.

The retina contains a large number of rods and a smaller number of cones. The cones serve for visual acuity and colour perception, while the rods are for vision in the dark, as in man. A shark will thus be able to distinguish a fish against dark background or a prey at the surface at night. It will do it all the better because it possesses a remarkable structure found only in nocturnal animals such as the cats: the *tapetum lucidum*.

Samuel Gruber of the University of Miami has been studying this structure for some years. The *tapetum lucidum* is a kind of mirror located behind the retina and composed of silvery platelets

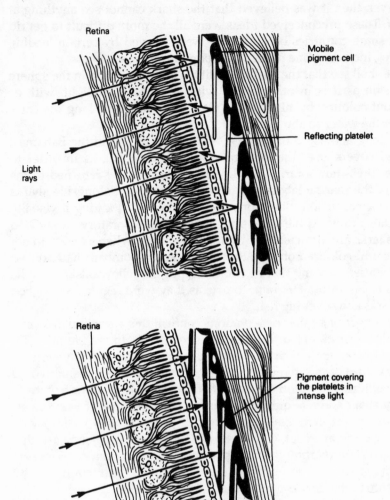

Retina

Mobile
pigment cell

Reflecting platelet

Light
rays

Retina

Pigment covering
the platelets in
intense light

A. In semi-darkness, the silvery platelets of the *tapetum lucidum*, located behind the retina, reflect the incident rays of light and so these pass through the retina a second time, considerably increasing the night-time sensitivity of the eye.

B. In intense light, the pigment cells slide along the platelets, preventing the light from being reflected back towards the retina. The latter is therefore stimulated only a single time, which is sufficient in normal light.

containing a pigment with a guanine-crystal base. At the base of these platelets are cells known as melanoblasts containing black pigment.

In normal lighting conditions (B), the melanoblasts slide in front of the platelets, which are thus rendered useless. In conditions of minimum lighting (at night or in ill-lit waters or in the daytime at great depth), the melanoblasts are drawn back to the base of the tapetum, and the light rays which have passed through the retina once are reflected by the silvery platelets and directed again towards the retina, which is thus stimulated a second time (A). The animals endowed with this faculty have eyes which shine at night, and as a general rule are nocturnal hunters (like the majority of sharks living in open water). This capacity to see at night has in particular been studied in the Lemon Shark both in the laboratory and in the sea – where it has been possible to examine the nocturnal hunting habits of the shark by radio telemetry tracking (using a probe placed on the animal), and scientists have been able to observe the shark arriving in hunting territories at night.

The capacity for night vision goes hand in hand throughout the animal world with an increase in the number of rods at the expense of cones, which more specifically permit colour vision. Hence the theory of some that sharks cannot distinguish colours. We have already mentioned the attraction of sharks for orange and yellow and their disinterest in black or blue.

Numerous theories have also been circulated about sharks' attraction to white. It has been said that sharks prefer white to dark colours, pale-coloured fish to dark ones, and make more attacks on white people than on darker-skinned races. It has also been said that, when they attack coloured people, sharks are more attracted by the palms and the soles, which are paler. On this last point, it is true that in certain regions of the Caribbean the black-skinned underwater fishermen still colour their palms and soles with black polish so as not to attract sharks, which seems to confirm at all events that these fishermen do not in the slightest feel immune from attacks because of their colour. I remember taking part in dynamite fishing in a small archipelago off Carthagena, Colombia in 1970. We had two dugout canoes, and, of the four black fishermen, one had no feet and another had only one hand. When I asked them if this was through mishandling dynamite, they replied wide-eyed that it was the "tiburons". I later noticed that many of the black fishermen had lost limbs, always through sharks, which were doubtless attracted by the shock wave of the dynamite, the blood and the vibrations sent out by the many wounded fish.

It has been said that dropping sheets of white paper into the water acts as a lure to sharks. In experiments carried out with both white and dark-coloured baits, observers have noted that there were many more attacks on the white ones than on the others, and from this it was deduced that bathers should wear dark swimming costumes rather than white ones. But the statistics are not altogether clear on this point, and it seems that the shark does not in fact discriminate in the matter. It attacks what is available, dark or coloured swimmers in the tropics, whites in the rest of the world.

With regard to movements around it, the shark's wide-ranging vision is an asset, particularly in seeking out and seizing escaping prey. As long as they are in a "food-seeking phase", many big sharks considered dangerous will detect movements around them and in general will move away from sudden movements such as arms being raised. On the other hand, once their "hunting phase" has begun, nothing will stop them and they ignore movements both above and below the surface, and will continue to attack and to feed even if a person strikes the water next to their head. This explains how very difficult it is to ward off a shark which is intent on attack, as we shall see in many examples.

The work of Gilbert and many others has shown that sharks are perfectly aware of violent or aggressive movements around them, above or below the surface. Beneath the surface all species realise that they are being watched. Great Whites, and to a lesser degree Tiger Sharks, are also aware if they are being watched from a small boat above the surface. There is now no doubt that sharks have the

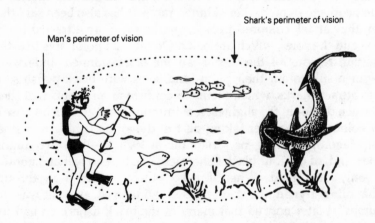

Man's perimeter of vision

Shark's perimeter of vision

In poor light, the shark's eyes are about ten times more sensitive than those of man.

capacity to see for a certain distance above the water. Some say they can see up to 2.5 or 3 metres around them, and quote the example of the sharks in Australia, near Darwin, which regularly leap into the air to seize those big flying squirrels known as "flying foxes" from tree branches.

As for close-range vision, here again sharks are highly impressive. Their eyes are large in relation to their body and much flatter than those of man, but that does not prevent them seeing just as well close up as they do at a distance. For close- range vision, it is the whole lens system that moves backards or forwards like a magnifying-glass, and not just the crystalline lens that alters its curvature as in the mammals.

Certain sharks, such as the Tiger, the Blue and the various *Carcharhinus* species (requiem sharks), possess a nictitating membrane or eyelid. When these sharks are excited or approaching a target, the membrane slips over the eye. The other species, which lack this membrane, such as the Mako and the Great White with their big black eyes, roll their eyes backwards before attacking, so that the white eyeball replaces the black pupil. Both these actions, the closing of the membrane and the retroversion of the eyeball, are symptoms which indicate excitement or signal a "feeding frenzy".

The visual system of the Great White Shark

We shall end this subject with the visual characteristics peculiar to the biggest predator of the seas, as observed by Gruber and Cohen (California, 1985).

As long ago as 1886, Schultze had formulated the theory of the existence of two forms of vision. The retina of nocturnal vertebrates was dominated by rod receptors, whereas that of diurnal species was composed of more cones. Animals with both cones and rods, i.e. with a composite retina, were reputed to have a protracted period of diurnal activity. Even now some people insist that sharks, and in particular the Great White, have poor eyesight on account of their having a lot of cones, even though their preferred habitat is in aquatic half-light, in caves, and they are apparently especially active at night.

Sharks have a composite retina, in other words they cannot only see during the day but can also distinguish colours. The diurnal activity of certain species which were thought to be nocturnal, such as the Lemon Shark or the Bonnethead, confirms the relationship between what we find in the retina and the way the animal's activity is divided during the course of the day.

Observations made on the retina of the Great White Shark, using an electron microscope to compare the percentages, the thicknesses, the distribution, the number of layers, and the pigmentation of different cells, have led to the following conclusions:

- *Carcharodon carcharias* has an eye perfectly adapted for good diurnal vision and good discrimination of colours

- it is less specialised for nocturnal vision than the Lemon Shark

- sight plays a decidedly more important role in its predominantly diurnal hunting life than in that of other, more nocturnal species.

One of the interesting observations on the histological level was the discovery in the centre of the retina, at the point where images are formed most sharply, the equivalent of the *area centralis* of higher vertebrates such as man, an area with a very clear predominance of cones. Again we also find the specialisation of the central retina for accurate, diurnal and colour vision, and of the peripheral retina for nocturnal vision.

The *tapetum lucidum* is more sophisticated in the Lemon Shark or the Blue Shark as its platelets are capable of changing angle so as to always remain perpendicular to the ray of light, and thus send more light back to the retina. In the Great White, the platelets are perpendicular to the retina over its whole surface, a slightly simpler arrangement which is in keeping with a preference for diurnal activity.

THE AMPULLAE OF LORENZINI: A UNIQUE ORGAN

The ampullae of Lorenzini are situated underneath the shark's nose and are endowed with properties of detection which are unique in the animal kingdom. They have several roles: detectors of variations in temperature and in vibrations, they also report variations in salinity, in contact pressure and infinitesimal variations in electric field. This last capacity could explain sharks' orientation methods as well as the migratory habits of some species.

Numerous experiments have shown that, even in completely opaque water, sharks were able to locate their prey, although it might be immobile and hidden in the sand. This cannot be explained either by vision, or by detection of movement, or by echolocation if the prey is, for instance, a skate buried in the sand.

It was Kalmijn who first discovered this hypersensitive detector positioned on the underside of the shark's head, well to the front and in front of the nostrils, which enables sharks to detect gradients of electrical potential of the order of a hundred-millionth of a volt per centimetre. This extraordinary sensitivity to electric fields which allows the shark to locate the weak electrical potential of its smallest

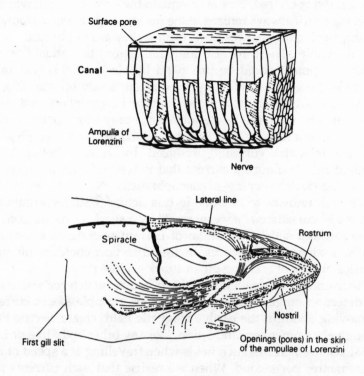

Electroreceptors of Lorenzini and (in black) branches of the lateral line. The external pores of the electroreceptors are indicated by black dots (from Kalmijn, 1978).

prey is again quite simply unique in the animal kingdom. This organ also probably takes the place of sight at the moment when the shark, having got to within a few tens of centimetres of its prey, very often covers its eye with the nictitating membrane or rolls its eyeball backwards.

Many experiments centring on the Lorenzini receptor were carried out, both in the laboratory and in the sea, with various species (such as Water Dogfish, Oceanic Blue Shark, swell sharks) from 1978 to 1985 by Kalmijn, Heyer, Tricas, Blonder and others. All these studies enable us to affirm that the elasmobranchs (rays, skates and sharks) detect their prey when the latter are in the immediate vicinity, and accurately seize them by using their power of electroreception. Luckily for the prey, their induced electric field diminishes very fast in the water the further they move away. Even in the case of a human body, which constitutes a "large prey", the gradient of potential is below the level of shark detection beyond 1 metre. Like all animals humans release a weak electromagnetic field, which becomes much stronger if it is injured. This may explain the precision with which an attacking shark always returns to the person it has already wounded, ignoring any uninjured rescuers, even if they are in physical contact with the victim. There are also some indications that attacks on man or his equipment could be provoked by electric fields that might resemble those of natural prey. One recent study on the sting ray (Blonder, 1985) has actually shown that the Lorenzini organ does not distinguish between the field emitted by a prey (shrimp, etc) and the same field emitted by a non-prey (sea-squirt, for example). The galvanic (electric) currents produced by contact between two different metals induce a current that is well within the perception zone of the electroreceptors of elasmobranchs.

Research remains to be done in this field, for such currents may perhaps be considered "supernormal" by a shark, or on the contrary could through sudden variation of current operate as a means of repulsing sharks. So research in this direction could result in an electrical means of warding off an attack at close range.

Other gradients of potential exist in nature, which are also within the detection range of elasmobranchs. For example, ocean currents, by moving about in the earth's magnetic field, create electric fields through electromagnetic induction, like any other body live or inert. The shark itself creates such fields when travelling at a speed of only 2 centimetres per second. When we realise that such currents have gradients of 0.05 to 0.5 volts/cm and that they are dependent on the direction of movement, it is evident that sharks have the necessary "equipment" to take their bearings in relation to these currents, in the manner of a standard electromagnetic compass. It has thus been

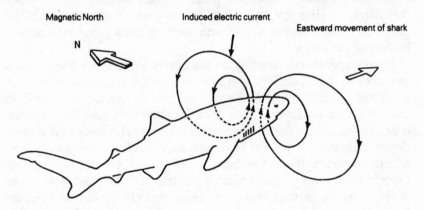

Magnetic North Induced electric current

Eastward movement of shark

N

A hypersensitive compass (from Kalmijn, 1978)

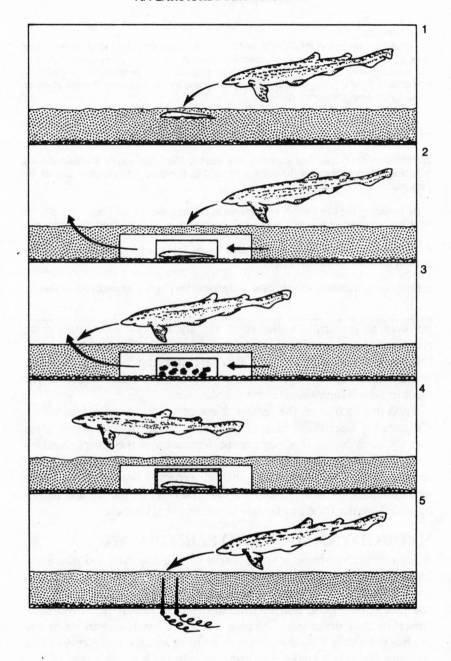

PREVIOUS PAGE:
Experiments conducted by Kalmijn in 1971, with a Lesser Spotted Dogfish, showing how a shark can detect a prey in absolute darkness or a prey completely buried in the sand. When we know that all sharks possess such faculties, it is easier to understand how it is impossible for man to escape any determined investigation by a shark in the neighbourhood.

1. Natural situation: the dogfish detects the sole (or the plaice or the dab) both by smell and by the electric field (1/25,000 of a microvolt per cm).

2. Enclosure that does not suppress the electric field but forces the odoriferous substances downstream of the current created in the sand. The dogfish detects the prey perfectly.

3. By replacing the live flatfish with pieces of whiting there is no longer any electric field, but the dogfish detects the odour where it comes out of the sand downstream.

4. The enclosure is completely insulated as regards electric field and smell; the dogfish does not notice anything.

5. Electrodes simulate an electric field in the sand: the dogfish immediately rushes in.

---

possible to confirm for the sting ray its ability to orientate with reference to the earth's magnetic flux. The great migrations observed for some species by John Casey are not the result of aimless wandering, but of well-oriented journeys to other regions, for reasons which themselves remain unknown.

I end this section on the remarkable sensory organs of sharks with a diagramatic description of an experiment made recently by Kalmijn (see previous page). This experiment demonstrates the complementary and supplementary nature of these organs and their sensitivity. It also shows that familiarity with sharks' sensory organs aids better understanding of their habits, and this understanding is the sole means by which we may one day be able to prevent shark attack.

HYDRODYNAMICS AND PERFORMANCE

A famous aircraft-builder was fond of saying that "a good plane is a plane that is nice to look at", alluding to the parallel between aerodynamic performance and aestheticism. In the same way we could say that a good marine predator is an animal with harmonious structure. Any diver who has seen a shark at least once in his or her life has been struck by the purity of its lines and the suppleness of its movements, which make it unquestionably the best endowed of fish for slow or rapid swimming, acceleration or pursuit, cruising or hunting. Here again the shark's adaptation to its environment is flawless, in the streamlining of its body, the size and the positioning of its fins, and the shape and proportion of its tail.

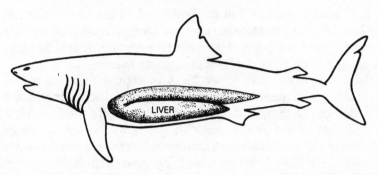

The Basking Shark as an example of a species that requires a huge liver in order to remain a long time just beneath the surface

If the basic principles of aerodynamism and hydrodynamism are the same, water is nevertheless eight hundred times denser than air, and the shark's propulsion necessitates a locomotive power which is much more considerable in equivalent weight than that of a bird of prey in the air.

AN ORGANISM FOR HYDRODYNAMISM

The first parameter to be set for any aquatic animal is that of its buoyancy, which must be variable at minimal energy costs. So the bony fish possess an inflatable swimbladder, and the sharks have acquired – during the course of evolution – an extremely sophisticated double adaptation.

The first adaptation transformed these bony fish into cartilaginous fish, replacing the solid bony frame with a cartilaginous skeleton. This development might seem regressive in comparison with the animal world as a whole were it not doubly beneficial:

- the gain in weight contributes to a reduced general density

- the gain in elasticity permits more supple and less "jerky" swimming, and faster acceleration owing to a greater capacity of movement.

The second adaptation is the acquisition of a very large liver, proportionately the biggest in the animal world, to the point where one wonders whether it is this factor that determines the overall length of the body.

This liver is all the more imposing the nearer to the surface the preferred habitat of the species in question is, and/or the more its pectoral fins are reduced in size. As in an aircraft, if the surface area of the "midships frame" of the body is doubled, the surface area of the fins will have to increase by more than twofold, just as the wing surface has to compensate the fuselage section. If the fins stay the same size, then it is the liver that must increase to hold more oil which is less dense than water. Thus, in Tiger Sharks, the increase in body size and diameter is not followed by a sufficient increase in the fins. A Tiger Shark of 2 metres in length has a fin area exactly equal to half that of a 4 metre shark of the same species, although the volume of the latter has more than doubled; it is the dimension of the liver that stabilises it by increasing by more than 200%. One might imagine that this progression would very quickly come to exceed the total capacity of the abdominal cavity, and there would be no room left for any other organ. It is this observation that leads to the belief that there is a maximum size for each species of shark. The only calculations made so far have been for Tiger Sharks, and we arrive at a theoretical size limit of 5 to 6 metres, which is perfectly corroborated by the entire stock of Tigers captured to date.

The liver of a big shark can reach 90 kilograms in weight, and it was well known, when sharks were hunted for their liver oil, that a 4 metre Tiger Shark had a liver that could contain up to 82 litres of the precious oil. The liver of the Blue Shark can be as much as 20% of the total body weight.

The other organs which are of lower density than water and thus contribute to the reduction in body density are the skin, the white muscles, the cartilaginous skeleton, and, in the Blue Shark, a light-weight jelly contained in the snout.

All this contributes towards the considerable reduction of the shark's density (which nevertheless remains heavier than water), with variations depending on the habitat. The Blue Shark (*Prionace glauca*) and the Spurdog or Spiny Dogfish (*Squalus acanthias*), which

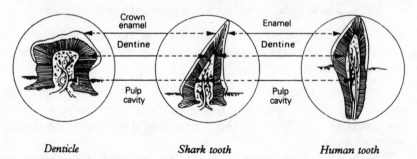

Denticle   *Shark tooth*   *Human tooth*

Histological analogy between the denticles and teeth. The pulp cavity contains nerves and vessels.

live in shallow water, have a weight in the water equivalent to 2.5% of their weight in the open air, while the Angel Shark (*Squatina squatina*), which moves about on the sea bottom, has a weight equivalent to 5%.

Irrespective of habitat, we might think that it would be advantageous for the fast-swimming species to have the lowest possible density, so as to reduce the area necessary for which the fins provide hydrostatic pressure, while at the same time reducing drag. However, this would be to ignore the fact that these fins must have a surface area large enough to allow adequate acceleration and manoeuvrability. Evolution has selected the best compromise between all these parameters to arrive at the highest-performance "submersible" imaginable.

The skin is one of the attributes that confers on the shark its remarkable hydrodynamism, to the point that the US Navy at one time took an interest in its mysterious texture, even envisaging covering an experimental submarine with it. I have explained elsewhere certain qualities of this skin, which incorporates a multitude of small modified teeth (the denticles) which are responsible for its considerable roughness, especially in the tract from tail to snout. Such a skin might seem incompatible with good hydrodynamism, and this has long intrigued researchers. In fact it seems that it is the particular alignment of these denticles which channels the flow of water and prevents it from moving away from the skin, thus producing laminate flow. This type of flow induces a considerably reduced drag in comparison to that of turbulent flow. It has moreover been suggested that this arrangement allows the shark to be silent in the water, an obvious advantage for a predator. It has also been observed that the denticles are smaller and more delicate in the faster swimming sharks (the pelagic species are faster than the benthic species living on the bottom of the ocean).

Sharks' fins work exactly like the control surfaces of an aircraft, which enable it to rotate around the three axes of roll, pitch and yaw.

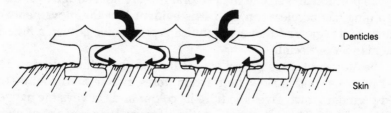

Diagram of the skin surface of a shark. The water passes between the sharp scales of the denticles towards the surface of the skin.

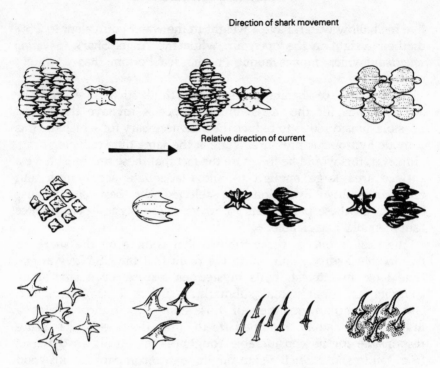

Some morphological examples of the denticles of sharks. The cutaneous denticles are as varied in their shape, their size and their arrangement to the extent that they are just as characteristic of any given species as the teeth. The discovery of such denticles in skin lesions can allow an attacker to be identified after the event. Despite their strange shapes, it seems that the denticles play a large part in the hydrodynamism of sharks.

The ailerons are replaced by the pectoral fins, the elevator by the pelvic fins, and the rudder by the upper lobe of the caudal fin. We could add that for stability the tailfin is replaced by the dorsal fin and the tailplane by the caudal keel. The fins have in addition a power of propulsion, of acceleration and of braking as a result of their axes of rotation which vary according to requirement, while a control surface of an aeroplane is only movable around a single fixed axis.

The position and size of the pectoral fins are not enough to explain lightning-fast acceleration, and it is evident that the movements of the caudal region add a very significant propulsive power, a little in the manner of a scull.

The motors of these propellers are the red and white muscles. The red muscles contain slow-contraction fibres, rich in myoglobin – the red pigment which gives the fibre its colour and conveys the oxygen to the muscle. These muscles are the ones for long periods of effort and activity, which necessitate an abundant inflow of oxygen.

The white muscles are predominantly white fibre and work anaerobically (without oxygen) and without myoglobin. These are the muscles used for brief, violent effort.

In the shark, the distribution of the two types of muscle is much less varied than in the higher mammals:

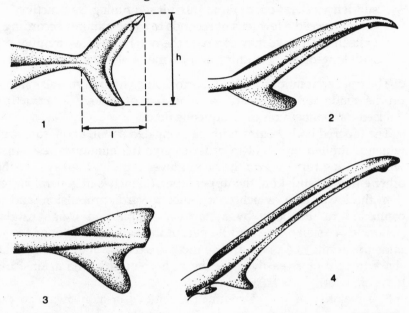

Principal types of shark tail fins.

1. Great White Shark: chiefly a coastal species, the Great White possesses a body and a tail comparable to those of the tunny. Slow cruising and sudden acceleration are secured by the tail fin, the two lobes of which are roughly symmetrical.

2. Tiger Shark: armed with a strongly heterocercal (upper lobe longer than lower) tail fin, it is the upper lobe that provides the main part of the locomotive power for slow swimming and of the thrust for acceleration.

3. Cookie-cutter Shark: armed with a large tail that is almost symmetrical, this shark chases squids and crustaceans and, using its labial suckers, clamps itself on to tunnies, dolphins or big sharks. With its spoon-shaped lower jaw, it cuts out cone-shaped plugs of flesh from its host. The luminosity of the tail also attracts various prey.

4. Thresher Shark: this shark of tropical and temperate seas pursues fishes and squids and stuns them with its upper caudal lobe, which is almost as long as the rest of the body.

The aspect ratio h/l (ratio of height of tail fin to its length) determines the hydrostatic thrust: the higher it is, the greater the thrust (as with the Great White and the Lamnid sharks in general).

- the red muscle forms a fine layer just beneath the skin, and works in the course of prolonged and slow swimming

- the white muscle is situated beneath the red muscle, utilises the anaerobic dissipation of the glycogen, and is made use of in sudden acceleration, or rapid pursuit swimming. Such activity cannot exceed a few tens of seconds to a few minutes according to its intensity (we may compare the two types of swimming with long-distance running and 100 metre or 400 metre sprints).

The final adaptations for high-performance hydrodynamism are a profile made for speed, and a general morphology of structure devised for manoeuvres in an aqueous fluid.

The tapered body begins with an elongated snout, and a fore area which is limited in height in order to give the minimum resistance when making turns where the head moves rapidly sideways. At the other end is a caudal fin, the upper lobe of which is in general larger than the lower lobe, which contributes towards propulsion, and is connected to the body by a narrow, supple and tough caudal peduncle or "stalk". The middle part of the body corresponds to the midships frame of a plane and so increases drag, but also allows the shark to turn more easily by acting as a pivot of inertia against the mass of water.

The majority of sharks swim in a sinuous fashion owing to the transverse waves released by the muscles. These waves are concentrated in particular in the region of the tail, the peduncle of which is narrow so as to diminish its drag as much as possible. The tail therefore supplies thrust which can be dissymetrical if the upper lobe is bigger, pushing the rear end of the animal downwards. This is compensated for by the pelvic fins, which create an upward force, and the flattened regions of the head then offer static resistance to the upward movement of the fore part of the body (it is interesting to note in fact that the very long upper caudal lobe of certain species is often accompanied by a more flattened head). Even without expert knowledge it is easy to distinguish a bottom-dwelling from a surface-dwelling shark. The upper lobe of the tail of the bottom-living shark is bigger than the lower lobe. Whereas sharks such as the Mako and the Great White have a symmetrical tail, indicating that they feed at the surface and at medium depths. The flat sharks are sharks of the sea bed, their shape allowing them to bury themselves in the sand.

A PREDATOR WITH TWO GEARS

The shark is capable of rapid speeds when it comes to hunting and chasing a definite prey, and of much slower cruising speeds when patrolling from one territory to another.

The cruising speed is not proportional to the animal's length, but of the order of only 1 to 4 km/h for sharks as a whole. This is a very low speed, lower than the normal walking speed of a man (6 km/h). It has been possible to measure this speed by direct observation, either underwater, from a boat or a helicopter, with a stopwatch over a known distance, or by telemetry. All studies put together, the following results emerge:

- Blue Shark: 1.3 km/h by day, 2.8 km/h at night

- Bull Shark (*C. leucas*): 2.5 km/h

- Spurdog, or Piked or Spiny Dogfish (*Squalus acanthias*): 1 km/h

- Great White Shark: 3.2 km/h

- Whale Shark (*Rhiniodon typus*) and Basking Shark (*Cetorhinus maximus*): 3 to 5 km/h.

Highest speeds are much harder to ascertain as one must be present at just the right time, but some observations give a good idea of the order of magnitude: the following figures all relate to one-off but verified observations:

- Blue Shark (60 cm): recorded average of 38 km/h, with peaks to 69 km/h

- Black-tip Reef Shark (*C. melanopterus*) struck on a line: 29 km/h

- Mako: 35 km/h.

VARIATIONS AND ADAPTIONS

The extraordinary longevity of the sharks over 350 million years has allowed them to adapt perfectly in accordance with their habits, in the sphere of hydrodynamism as in that of sensorial detection.

The Lamnidae family (including the Great White, the makos and the porbeagles) has adapted to a pelagic environment which called for long-distance movement, and therefore a high cruising speed, and high-speed chases, which in turn call for considerable power. These requirements explain the general torpedo shape of these sharks, and their sizeable "midships frame", the upwards extension of their tails, the large dimensions of their liver and, a specific adaptation of this group, their body temperature which is 5 to 10 degrees above the ambient temperature. This last characteristic enables better output by their highly developed red muscles which, in contrast to other species, are embedded in the body around the backbone. This muscle tissue is closely joined to a network of capillaries with which it exchanges the heat produced by muscular activity.

A unique experiment was successfully made off New York, where a 4.6 metre Great White Shark was tracked uninterruptedly for 80 hours using acoustic telemetry. In this way it was possible to record the depth at which it moved, the temperature of its muscles and that of the surrounding water. Its average speed was 3.2 km/h and most of the time the animal remained in the thermocline (the boundary between cold water and warm water). Its muscle temperature remained 3 to 5 degrees higher than that of the surrounding environment, which is exceptional for a homeotherm. In Great Whites, high-speed swimming is aided by the existence of a precaudal keel which provides stabilisation.

In contrast to these very active species, some more sluggish species such as the Carpet Shark and the dogfishes (Scyliorhinidae) keep to the immediate proximity of the sea bed. A large head and a small tail, a low percentage of red muscles (8%), and a high density (4.7% greater than that of water) are characteristics of species which move about very little and very slowly and which live in the depths. Their large fins allow them good manoeuvrability on the ocean floor, and are not shaped for speed. These species move like eels, with wide swings of the front and the rear of their body, and their tails contribute only a little to hydrodynamic thrust.

Certain species have feeding habits requiring them to have neutral buoyancy (density 1). This is the case for the Goblin Shark, the Frilled Shark and a large number of the dogfish (Squaliformes) which live in deep seas. Food being scarce in the ocean depths, they must be capable of moving rapidly upwards towards other prey. They manage this effectively thanks to their very large livers which make up up to 25% of their body weight and contain up to 90% oil made up of a product of very low density, squalene. The density of the squalene varies with depth. Thanks to this constitution, they can come up more quickly than their prey, in particular more quickly than the bony fishes whose swimbladder is less efficient in performance.

Living much nearer the surface, since they feed on plankton, the biggest of the sharks, the Whale Shark and the Basking Shark, also have a neutral buoyancy owing to their enormous livers filled with squalene. This allows them to swim at 3 to 5 km/h in the plankton layers. All these sharks are sought after for the squalene in their livers, as it is used in cosmetics.

The hammerheads are one of the strangest shaped sharks. One of the functions of their very broad head is to operate in the manner of a wing, inducing an upward hydrodynamic thrust at the front, which permits sudden vertical movements. When changing direction, this flatness of the head offers minimal drag to the lateral movement. It is as a result of this very peculiar anatomy that the hammerhead can catch its preferred prey of squids, animals which move about reactively and are therefore extremely mobile in all directions. In the hammerhead *Eusphyra blochii*, the width of the head reaches up to 50% of the body length.

These few examples help us understand why it is so difficult to escape certain sharks, and show the extreme handicap of the human being underwater in comparison with these animals.

A final detail concerning movement, and one which is a great disadvantage to sharks is their inability to move backwards. The musculature and morphology of the fins are designed for forward motion only, which explains the rapid suffocation of sharks when caught in nets, as they are unable to free themselves other than by swimming forwards.

RESPIRATION

The cruising speed of swimming can be maintained only if the oxygen supply to the muscles is correct. Oxygen is extracted from the water in the region of the gills. Sharks' gills are made up of cartilaginous arches supporting a series of gill rakers which are perpendicular to them. These rakers in turn support secondary bars which are also perpendicular and direct the flow of water taken in through the mouth in the opposite direction to the blood flow. Exchanges of oxygen and carbon dioxide between blood and water therefore take place all the more quickly.

The water taken in through the mouth is discharged through five gill slits in the majority of sharks (six or seven in the most primitive species, though nobody knows why).

The process can be accelerated in the fastest-swimming species through the "ramjet principle", in other words the exchange happens better the faster the shark moves. Lamnidae sharks have to keep moving to ensure this process which means that if they stop they die from suffocation (this accounts for the majority of the big predators caught in beach-defence nets being brought up dead).

By contrast, some bottom-dwelling species (dogfish, carpet sharks etc), with no regular movement, pump water by rhythmically contracting the muscles that control the inlet and outlet valves of the gill system.

Between these two extremes, certain species such as the Sand or Sand Tiger Shark, also known as the Grey Nurse Shark (*Eugomphodus taurus*), are able to make use of both pumping and ramjet, achieving a real saving in energy.

To ensure a good transfer of oxygen from the gills to the muscles and other organs, a maximum percentage of red corpuscles in the blood (haematocrit) is needed. In the benthic species living at great depths, the haematocrit is of the order of 15%, while it reaches 35% in the Lamnidae family, the most active species.

A final parameter for good oxygen transfer is the cardiac pump. Unlike in mammals, there is no relationship between the heart size and the body weight or the activity level of the animal. The ratio of heart weight to body weight remains constant, of the order of 0.1% in all species, no matter how active. Only the Lamnid sharks, including the Great White and the Mako, have a ratio reaching 0.2% to 0.3% (which gives a heart of only 4 kilos in weight for a monster of 2 tonnes). This means that even these very large predators do not exhibit exceptional stamina in prolonged exertion, even though they are capable of enormous exertions in short bursts.

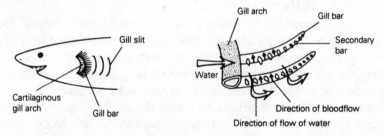

Structure of the gills of a shark showing the counterflows of water and blood

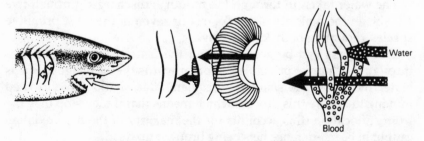

Water enters through the mouth and leaves again by the gills, where the blood circulating in the capillaries gets charged with oxygen dissolved in the water.

The spiracles are external openings of the respiratory apparatus of certain rays and sharks which, living on the bottom, make use of this in preference to their gills.

When big sharks are caught for aquaria, they are forcibly kept in an artificial current in order to "resuscitate" them, and to enable them to get over the shock of capture and transportation.

As already mentioned, when a shark is caught up in a beach-defence net – which exist in Australia and South Africa – its forced immobilisation leads to a rapid fatal asphyxiation. At the atomic power station in the Cape, sharks are sometimes caught in the suction gratings, the diameter of which is about two metres. However, in these cases they do not die, as a current of water is artificially maintained in their gills by the suction. On the other hand, when special clasps are used to haul them to the surface, they die very quickly, their backs breaking under their own weight.

4

PECULIARITIES AND CURIOSITIES

THE "GREAT WHITE DEATH"

IT IS TO ITS BEST-KNOWN representative that the shark world owes its association with the idea of "white death". The Great White Shark, *Carcharodon carcharias*, has always in fact been the most feared of the sharks, but also the most misunderstood as regards its physiology, and the least observed in its natural environment (up to 1965, no camera had ever filmed it underwater). The name "white death" comes from the particularly brilliant snowy whiteness of the whole of the underside of these sharks. While the upperside has nothing distinctive about it, varying from indigo through brown or grey to green, the white colour is immaculate in *Carcharodon carcharias*, producing a very well-marked and characteristic, sharp contrast with the upperside. It was the whalers who first used this name for the monsters that shamelessly came and tore hundreds of kilos of flesh from the whales they were towing.

Victim of a very large number of preconceived ideas, like all its congeners, there are nevertheless an increasing number of definite known facts about the Great White Shark. It is undeniable that it remains the most dangerous of the marine animals, and that it is found in temperate as well as tropical waters, occasionally reaching very high latitudes. It is not exceptional in the Mediterranean, where some fatal attacks in the Adriatic are attributable to it. It has been seen in Corsica, at Propriano, and it could be responsible for some of the unexplained disappearances of swimmers.

One of the features of this shark is its preference for depths of over thirty metres. This might explain why it is more frequently observed on the Pacific coast of America than on the Atlantic coast. The 30 metre isobath is several kilometres out from the coasts of the Atlantic, whereas, on the Pacific coast, the sea bottom plunges very steeply and depths of a hundred metres are frequently found at the foot of cliffs. Thus on the west coast of the United States, there are three or four attacks on swimmers or surfers every year, while on the east coast, at identical latitudes and in comparable temperatures, they remain very rare.

The teeth of the Great White are equally sharp-edged in upper and lower jaws, giving rise to very neat, linear cuts through the toughest skin. Fossilised teeth have been found, belonging to a species that disappeared ten million years ago, whose biggest teeth reached a size of 18 centimetres, which, by extrapolation, gives a conceivable jaw diameter of about 2 metres and a monster with a length of around 20 metres. These teeth were more finely notched than those of the present-day Great White.

The maximum size that may be attained is 9 to 10 metres for a measured specimen, even though certain statements that are difficult to verify have cited even greater sizes. If we note that the Megamouth Shark was not discovered until 1976, it is conceivable that some colossal specimens of Great Whites exist in the depths. In August 1986, off New York, a 7 metre specimen was watched for a long period.

It seems that the Great White Shark may show a certain tendency towards territoriality, and that it returns to the same areas year after year. This could explain series of attacks on one and the same coast, spread out over several years. Tagged individuals have been found

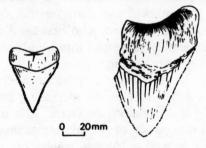

0   20mm

Relative tooth sizes of the modern *C. carcharias* (4-metre Great White Shark) and of fossil *C. megalodon* (size estimated at 12 metres)

again six months later only one or two kilometres from the place where first captured.

I have mentioned elsewhere the learning ability of sharks, and the Great White appears to be endowed with a memory superior to that of the average fish. Observers report that a Great White had violently bitten the steel cage in which they were protected, and that it had lost several teeth in the process. On a second encounter in the same circumstances, the same shark hesitated for a long time before biting, only finally deciding to resist the temptation, as if it remembered that it would break its teeth on the cage.

For scientific observation of Great Whites, the only method is to have a shark cage at one's disposal on the one hand and, on the other, to know how to lure them with bait. For Ref Dean, regarded as one of the greatest shark fishermen, the ideal bait consists of a mixture of tunny flesh laced with fish blood and fish oil. It is clear that observations obtained in these conditions are only fragmentary, even though they are in line with the principal ethological motivation of the shark: feeding. Apart from this feeding behaviour, we do not know exactly what the Great White Shark does, and thus speculation sometimes proceeds at a fair pace.

We have been able to show, however, that the members of Lamnidae family (which includes the Great White) have temperatures higher than their surroundings, which allows them improved performance (see Hydrodynamics and performance).

Among the characteristics peculiar to the Great White, is that it is the only fish to bring its head out of the water in order to observe in the open air what is going on around it (see photographic section).

INNOCENT GIANTS

Even though they present no danger to man, I shall say a few words on the two largest living fish, for they are sharks: the Whale Shark and the Basking Shark.

Capable of reaching 18 metres in length and more than 40 tonnes in weight, the Whale Shark (*Rhiniodon typus*) feeds on plankton, small fish and crabs, which get engulfed in its wide- open mouth (average diameter 1.15 m). This is provided with filters which prevent it from taking anything larger in size than a small tunny. Usually feeding on the surface or just below it, it often holds itself in a vertical position with its mouth flush with the surface. This explains some collisions or shipwrecks. In 1934, for example, the ship *Maurganui* struck a Whale Shark in the South Pacific: the fish was impaled on the ship's stem, 4.6 metres of it on one side and 12.2 metres on the other. The *Armadale Castle* suffered the same mishap in 1905 in the Atlantic:

THE JAWS OF DEATH

another 17 metre mastodon was rammed by the ship's stem and remained stuck for a quarter of an hour, propelled through the water by the ship, which was fortunately solid and sturdy.

When not at the surface, the Whale Shark lives in the deep ocean currents where it allows itself to be carried lazily at a speed of two or three knots. When it leaves the surface, it does so in the manner of whales, diving vertically, as the two following anecdotes demonstrate. The first was reported by Grey, a great fishing fanatic, who had an encounter with a 15 metre Whale Shark off the coast of Mexico. When the fishermen managed to harpoon it, the mastodon dived, dragging with it more than 500 metres of rope, before ridding itself of the harpoon (its skin can be up to 23 centimetres thick). The second anecdote was related by another game fisherman who harpooned a 12.8 metre Whale Shark in the Gulf of California in 1938. The harpoon was connected by a rope to a large empty petrol drum. As soon as it was hit, the animal dived to the depths and when it reappeared a quarter of an hour later, the drum was completely crushed from the pressure it had undergone. Renewed attempts to harpoon it failed, for, once wounded, the Whale Shark is capable of contracting its muscles, making them as tough and resistant as an armour-plated ship.

The peaceful and passive nature of this animal is such that divers have been able to ride it without its doing anything whatever to break free. At the very most it exhibits a slight curiosity with regard to certain divers, coming to inspect them and then leaving again.

The biggest Whale Shark measured reached 17.98 metres and weighed nearly 43 tonnes, and was caught in a trap on the eastern coasts of the Gulf of Siam in 1919 in 15 metres of water.

According to Gudger, another specimen is said to have established itself, around 1890, in the vicinity of San Juan in Porto Rico and to have stayed there for some years. When it died, its body was hoisted on to the bank, and it was found to measure 20.5 metres.

If you happen one day to come across this mastodon, do not be alarmed. Approach it if you wish and do not be afraid of being mistaken, for it is easy to recognise: grey-brown on the back and yellowish-white on the flanks and belly, it has narrow circular bands of yellow and white along its entire length.

The Basking Shark (*Cetorhinus maximus*) is not quite as big (a 13.71 metre specimen was observed in Norway, and another of 10.93 metres off Brighton in Sussex), but just as inoffensive, and also feeds on plankton. We have mentioned elsewhere that its enormous liver gives it neutral buoyancy. Much less cosmopolitan than the Whale Shark, it is found on all the coasts of Europe, North America and

South Australia, as well as on the coasts of China, Chile and New Zealand. Occasionally it moves about in groups of 100 to 150, and its carcase is often found stranded on a beach. This is a fish that may swim in single file with several following each other very closely, thus giving the illusion of a single enormous sea serpent, and is the possible origin of many of the fears of the past.

In 1976, when the crew of an oceanographic vessel of the American Navy raised its sea-anchor off Hawaii, they were amazed to discover an enormous shark jammed in its cables. The animal weighed 726 kilos for its 4.46 metres, but looked like no other shark known, having an enormous mouth equipped with minute teeth. In its stomach were found shrimps of the open ocean and nothing else. The structure of its gills confirmed that this unfamiliar fish was a filter-feeder, sifting food particles suspended in the water. The internal part of the gills was covered with crossed filaments like a fine-mesh net, allowing the animal to feed by moving forwards with its mouth wide open. This is the third shark that we find feeding thus in the manner of whales. Although this one also possessed an additional trap to capture its prey: set out all around its mouth are rows of luminescent organs, glowing like glow worms, and intended to attract the small shrimps present in the plankton.

The scientific community, at first incredulous, christened this extraordinary new specimen "Megamouth", and classified it in a separate family of which it remains the sole member, the Megachasmidae. A second specimen was discovered in 1984 off the island of Catalina in California, and a third in 1990, proving, if need had been, that the oceans have not given up all their secrets to us.

THE TIGER SHARK: THE "HYENA OF THE SEAS"

The interrelation between the sharks' habitat and their method of feeding is obvious, and the level of danger a shark presents to man depends very largely on its feeding habits. The Tiger Shark is particularly dangerous because, apart from its size and its spectacular jaws, it is also an opportunist when it comes to feeding, and when hungry swallows everything it comes across. It is certainly the least specialised shark in this sphere, and a scientist was able to say of it that it was a "dustbin with fins". Both predator and carrion-eater, it eats a wide variety of bony fish, including tarpon, moray eels, grey mullet, sole, etc. But it also eats other sharks such as sawsharks, the Grey Reef Shark, hammerheads and even other Tiger Sharks. It has the knack of attacking fish or its own congeners caught on hooks or in nets, to the point where it poses a serious problem for tropical fisheries. Fortunately, in its impetuous voracity, it often gets caught itself on the hook of its victim once it has swallowed it. It is the Tiger Shark that devours most marine reptiles, in particular turtles, and it is one of the major predators of sea snakes. It swallows the marine iguanas of the Galapagos – a large green iguana has been found whole in the stomach of Tiger Shark. Nor are birds missing from its menu: frigatebirds, cormorants, pelicans, as well as all the migratory landbirds which are unfortunate enough to alight at sea, are eaten. The marine mammals taken by the Tiger Shark include the sea lion, seals with or without their skin, dolphins and dead whales. Nor do crustaceans and invertebrates escape the Tiger Shark, which gorges itself on crayfish, crabs, octopus, etc. Any bodies of terrestrial animals are all acceptable: birds, chickens, rats, pigs, sheep, dogs, hyenas, monkeys and humans. All have been found in the stomachs of Tiger Sharks, some had clearly been swallowed when already dead – others were swallowed alive.

This shark is also well known for swallowing an incredible number of man-made objects: leather coats, bundles of wool or cotton, silk, pens, plastic bags, small cans, bottles, pieces of metal, etc. It is clear that nothing in a diver's dress can rebuff it, neither the neoprene wetsuit, nor the air bottle, nor the monitoring equipment with metallic reflecting surfaces, nor of course the diver himself.

DISPLAY SWIMMING OF THE GREY REEF SHARK
Quite a number of species live in the lagoons and on the coastal reefs of the Pacific, but there are only four of them known to attack in a relatively regular fashion, at least when provoked, excluding of course all the other pelagic species that regularly come to feed near the coasts. These species are, in decreasing order of danger:

- the Grey Reef Shark;
- the Silvertip
- the White-tip Reef Shark
- the Black-tip Reef Shark or Blackfin.

The last two species are, along with the Grey Reef Shark, the most widely distributed in the lagoons, but it is the Grey Reef that exhibits by far the most violent aggressiveness. This shark is the only one to have a threat posture, a scientific study of which was made in 1986 by Nelson and Johnson, which is called "display swimming" or "exhibition swimming". This threat posture warns that an attack is imminent and so must be known by divers or swimmers likely to observe it.

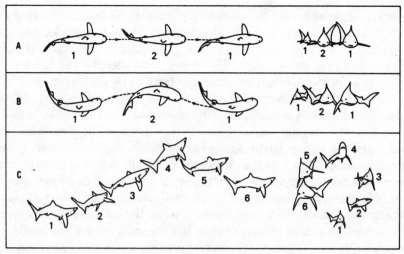

Comparison between normal swimming and display swimming in the Grey Reef Shark (from Johnson and Nelson, 1973):

A. Normal swimming. B. Display with laterally exaggerated swimming. C. Display with rolls (1-2-3) and spiral loops (1-6). Although the rolls are similar to the initial phases of the spiral loops, they are distinguished by the fact that the shark returns to a display posture in a level position, without describing the downwards path of the spiral loop. The display is all the more imposing the more the shark's escape route is impeded.

The first observed instance of such an attack was back in 1961, when Church, who was free diving at Wake Island, passed a Grey Reef Shark which then began to contort its body in all directions in erratic swimming movements, turning and half-rolling, bending itself towards the front then backwards, before suddenly uncoiling itself towards the diver, and inflicting two severe bites on his arm.

The same happened in 1967 to Fellows, who observed swimming with "exaggerated" movements immediately followed by a very rapid charge, which Fellows succeeded in dodging the first time and beat off the second time by striking the shark.

At Enewetak, in 1978, two free-divers found themselves at the bottom of a lagoon, at which point one of them took a flash photo of a Grey Reef Shark in full display swimming. The immediate effect of this was to trigger an attack. One of the divers was seriously bitten on the arm while the other was circled by the shark, which swam around him several times before tearing away one of his flippers and then biting him on the hand.

In 1975, a number of attacks were made on a small submersible, occupied by two people, which had been specially designed for following these sharks and observing their reactions. One of the sharks, pursued by the submersible, started its display swimming at a respectful distance from the contraption, then at a stroke turned around and made an "explosive" charge at the perspex dome, which was two centimetres thick. The dome was deeply cut, and the damage inflicted as much by the upper teeth as by the lower ones.

Convinced that observation of these early preliminaries to attack could be of interest in understanding shark attacks as a whole, Nelson and Johnson started on a series of observations at Enewetak in Micronesia, using a small submersible which protected them from direct attacks. In this way they triggered off 57 attacks, 38 of them by the Grey Reef Shark, especially when the latter was cut off or cornered in a twisting channel of the sea bed or the reef, and all preceded by display swimming. Attacks were very frequent when the submersible followed the animal in each of its movements, but we shall see that the machine was regarded as a potential predator rather than as a competitor (social or territorial) and on no account as a prey.

The attacks by the Grey Reef Shark can be broken down into two general categories: attacks during feeding periods and attacks outside feeding periods.

On the occasion of attacks in the first category, the shark is greatly attracted by stimuli such as sounds and odours emitted by a freshly caught fish, and there is no hint of display swimming, particularly if there is competition with other sharks (as in the "feeding frenzy" for example). This means that an undersea hunter carrying fish on or behind him will have no cause to believe himself safe just because he does not observe any display swimming.

Attacks in the second category, outside feeding periods, often go no further than the outline stage, such as intimidation charges (abortive attack), and consist of a sudden acceleration towards the target, this time always preceded by a demonstration of display swimming lasting about thirty seconds. This display must be made known by all those who frequent the lagoons, for it always announces an imminent attack by a Grey Reef Shark. This attack is not automatic if the threat felt by the shark is not aggravated (by a diver who simply passes by, for example).

The small submersible used weighed 100 kilos, was 2.5 metres long and was covered at the front by a perspex dome allowing very good visibility. Its mobility was provided by a small propeller, three electric motors and two sorts of "pectoral fins" at the front, enabling it to rotate 360° on its own length. When the Grey Reef Shark was followed by the submersible, three reactions took place: either (1) the shark would speed up and outdistance the submersible; or (2) it began a sort of "carousel" by turning in time with the submersible; or

(3) it stopped and displayed, intensifying the display as the machine got nearer. It was in this last event that attack most often occurred, at the point when the submersible approached to a critical distance of 2 to 3 metres. Just before the attack, the shark was often very tense, holding itself in a contorted posture, the fore part of its body almost immobilised, which caused it to sink slightly on account of its density being higher than that of the water. For this reason, attacks were very often launched from beneath the submersible. They were directed at the front part, and at the perspex dome in particular, but also at the stabilisers, the engines and even the polycarbonate propeller, which was destroyed. Attack, always very swift, occurred after 5 to 20 seconds of pursuit, and the films show that it never lasted more than a third of a second.

In two attacks only, the shark bit a second time, otherwise usually moving off or remaining in the vicinity, sometimes continuing its

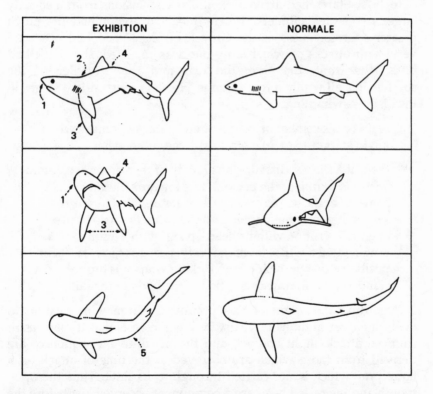

Comparison between details of posture in normal swimming and those in display swimming in the Grey Reef Shark. Explanation of arrows: 1 Raising of snout. 2 Resultant angle between cartilaginous cranium and vertebral column. 3 Lowering of pectoral fins. 4 Arching of back. 5 Bending of body.

display swimming. The single bite, as is often observed from large predators such as the Great White Shark, is perhaps only an intimidation bite.

It was noted that females or males attacked in the same fashion, that they were of normal adult size (1.4 to 1.7 metres) and that the specimens of maximum size did not attack more. The small ones (0.8 metres) on the other hand fled the most often without display swimming. Solitary individuals were more inclined to attack, as well as those that were on the bottom, having no escape route to the depths.

The most timid demonstrations were seen in circumstances which perhaps presented something new to the shark, involving approach out of curiosity without obligatory conflict: namely the presence of a non-threatening diver, a diver entering the water by dropping from the side of a boat, a boat casting anchor, and transmission tests of low-frequency sounds.

In 1975, Starck had arrived at similar conclusions from a slightly larger two-seater submersible. He, too, had ascertained the much greater aggressiveness of the Grey Reef Shark, and had observed very swift attacks of which only one was repeated: the shark had bitten the front, the side, the rear and the propeller of the submersible all within a few seconds. He had observed two stages in the display swimming:

- at first a movement of exaggerated lateral swimming which might start at 15 or 20 metres from the submersible

- then, at a shorter distance, an arching of the vertebral column, with a lowering of the pectoral fins towards the vertical. In some cases the display was so exaggerated that forward swimming was interrupted and the animal remained in a diagonal position with its head up and with a highly exaggerated angulation of the body. Perhaps the very "bent" position stops the shark from sinking when it is brought to a standstill, by an upwards action of the body or the tail.

Regarding the origin of this peculiar swimming in relation to evolution over millions of years, Barlow suggests that it may result from an attack/flight conflict, and that its different sequences are derived from those which are observed at the time of attack on a large prey which is not carried through to its limit. Thus the open mouth and the raised head are a component observed just before the bite, and the lowered pectoral fins are noted during biting to give some stability to the shark's body while it tears up its prey by violent sideways shakes of the head.

The Whale Shark is the biggest of all fish (up to 18 m) and the most harmless of the sharks. Shown here accompanied by several remoras. *(K. Amsler/JACANA)*

With its long pointed snout, big triangular teeth, long gills and immaculate white belly, the largest predator of the seas is also one of the most beautiful (*Carcharodon carcharias*: the Great White). *(D.R.)*

The Tiger Shark has a short blunted snout, big notched teeth and is often striped on the flanks. It is almost as dangerous as the Great White, and even more elegant. *(D.R.)*

The Great White is the only fish that raises its head out of the water to inspect the surrounding airspace. The Orca (Killer Whale) does the same by leaping. Both feed on seals and sea lions. *(D.R.)*

The Great White often prowls at the surface, its fin projecting, but it attacks most often from the depths, emerging at great speed, to surprise its victims and hurl them above the water. *(K. Amsler/JACANA)*

The jaws of adult sharks 4 m or more in length have a power of up to 5 tonnes / cm² (as against 220 kg for a 70 kg or 11 stone man). Even a steel cage cannot stand up to such pressures. *(M. Pignieres/SIPA)*

Sharks alone in the animal world possess several rows of reserve teeth (this Tiger Shark from the Persian Gulf has five). *(K. Amsler/JACANA)*

Many sharks, such as this Spotted Ragged Tooth (or Sand Tiger), are armed with very long, pointed lower teeth, curved backwards like fishhooks. Once prey is impaled on these teeth, the upper jaw comes sawing into the flesh. *(M. Pignieres/SIPA)*

During a fishing competition in Australia, a Great White came and swallowed whole a 2.10 m Dusky Shark of 80 kg which was hanging on the stern. This is photographic proof that a shark can swallow a man whole. *(D.R)*

The Grey Reef Shark will not hesitate to attack divers carrying wounded fish on them. *(K. Amsler/JACANA)*

Shark fishing is increasing – their fins are highly prized in Asian cuisine. *(X.M.C)*

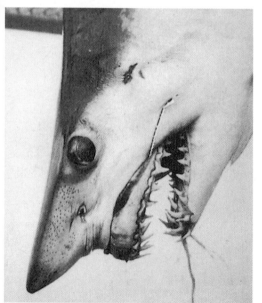

(ABOVE) The pores of the lateral line and of the ampullae of Lorenzini are particularly visible on the snout of this young shark.
(RIGHT) Attacks are increasingly being targeted on surfers, whose silhouettes resemble those of the pinnipeds (seals, sea lions etc) on which the big predators feed. *(D.R./NSB/X.M.)*

At best, the damage caused is only material. *(NSB/X.M.)*

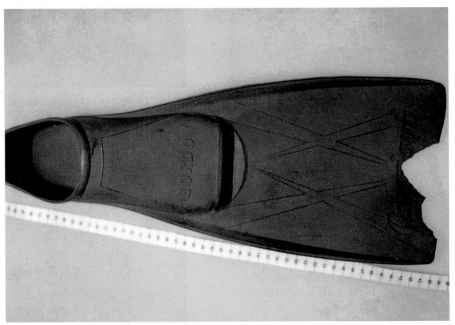

This swimmer was particularly lucky that his flippers were not to the liking of the shark that was following him. *(NSB/X.M.)*

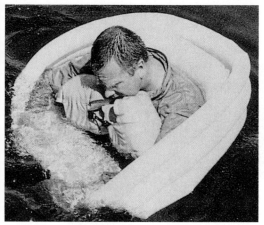

The Johnson shark bag is an excellent passive means of protection, lightweight and unencumbering. The shark sees only a shapeless and motionless dark mass which does not release any odours. (ABOVE) US Navy experimentation. (RIGHT) Prototype of the Natal Shark Board in Durban. (US Navy/X.M.C.)

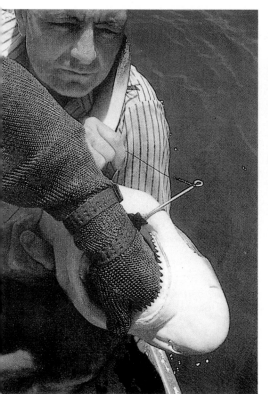

Another passive and appealing means is the mesh suit (here being tested by Ron and Valerie Taylor with Blue Sharks of middling size). This effective defence against penetration by the teeth is useless against the crushing jaws of a big shark. (R. and V. Taylor)

The investigation following an attack begins with the examination of objects destroyed by the shark. Here Beulah Davis (director of the Natal Shark Board) examines a victim's surfboard. *(Denize/GAMMA)*

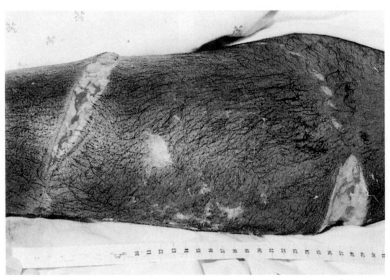

The diameter of the wounds, the shape of their shredded internal slopes, the depth of the bites, whether single or multiple, etc, are all examined. *(NSB/X.M.)*

The cage is the most effective of the passive means of protection for scientific study, photography, etc, although it is still inadequate against 6 m monsters weighing 2,000 kg. *(D.R.)*

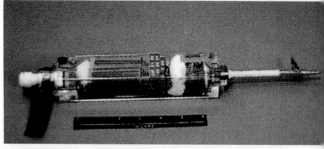

The CO_2 dart can be effective in unbalancing the animal or, by luck, bringing about a gaseous embolism. *(Johnson)*

This big Nurse Shark has been killed by a harpoon with an exploding head, an excellent active means of close-range protection. *(D.R.)*

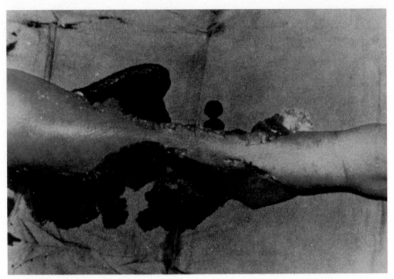

The extent, the depth and the force of the attack are recorded (here a heavily mutilated leg). *(NSB/X.M.)*

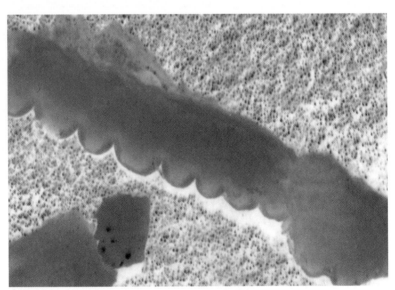

Tooth fragments found in the wounds or bones are examined under the microscope or the scanner, and can indicate clearly the species responsible for the attack (here a tooth fragment of a Great White Shark found in a wound during surgery). *(NSB/X.M.)*

Here, at the Natal Shark Board, the case of Sudash Sarjoo, attacked on 20 January 1989 at Isipimgo, is being studied. *(M. Pignieres/SIPA)*

Every morning, 43 boats of the NSB put out at dawn from 43 beaches of Natal province. Their task is to raise the 340 nets set beyond the barrier to prevent sharks from coming onto the coast. *(Photos NSB/X.M.)*

Sharks very quickly die of asphyxiation once they can no longer move forward. Those sharks brought up alive are released.
(Photos NSB/X.M.)

All the sharks that die in the nets are tagged and frozen before being autopsied for scientific purposes. (RIGHT) One day's catch during the 1985 "Sardine Run". The only days when the nets are not set is the time of the "Run" (BELOW), which brings millions of sardines inshore every year.

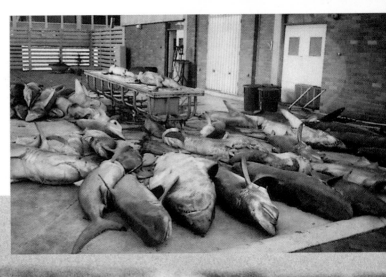

Too many together to be able to breathe, they die of asphyxiation and accumulate in great heaps on the beaches. *(NSB/X.M.)*

Autopsies are carried out in public three times a week, to inform the population. *(NSB/X.M.)*

The nets sometimes collapse during storms and the beaches are put out of bounds while divers work on them. The predators can then reclaim their domain. *(NSB/X.M.)*

Starck also noticed that it was not the ones which were the most demonstrative that attacked the most often, but quite the contrary. He also came across an episode in which a small shark was so excessive in its display swimming that it was destabilised by it, but it never attacked, even when the submersible bumped into it. This could be a question of a powerful inhibition against attacking, a conflict between the desire to attack and the "fear" of so doing.

This inhibition of action might remind us of stress in human beings, if we bear in mind that stress and inhibition are intimately linked. Stress in man manifests itself in numerous objective signs (muscle spasms, shaking, sweats, twitching, rapid heartbeats, etc) of which the exaggerated display swimming could be the equivalent when it does not lead to attack. Perhaps Starck's submersible, being bigger and more imposing, would more easily give rise to an abortive attack, while that of Nelson and Johnson, resembling a shark more in its shape and behaviour, could appear more like a real predator or a serious competitor and so trigger threat followed by a genuine attack.

A general principle justifying attack or otherwise for vertebrates is the degree of surprise caused by what they come across in their territory in relation to what they expect: a Grey Reef Shark can expect to meet another shark in its usual reefs, not shiny steel monsters that pursue it.

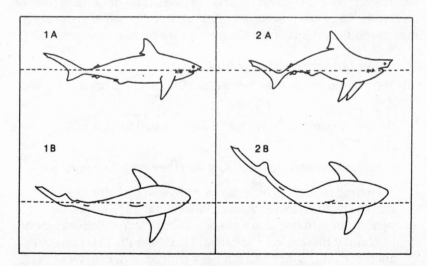

Less accentuated variant of display swimming: The aggressive swimming of the Reef Whaler. 1A and 1B: Reef Whaler in normal slow, horizontal cruising. 2A and 2B: Reef Whaler in phase of aggressive swimming: 2A back arched, 2B violent sideways swings.

In 1981, Nelson noted in other species a few signs corresponding to the early stages of the Grey Reef Shark's display swimming. For example, the exaggerated lateral swimming, with head swinging from one side and to the other, was observed in the Galapagos Shark in the Cocos Islands in the East Pacific. This raises the hypothesis that this posture could help the shark to maintain visual contact with objects situated directly behind it, a diver for instance (fish have a very wide field of vision compared with that of man, almost 180°).

The arched position with snout raised was also found in the Bonnethead (*Sphyrna tiburo*) but without the exaggerated swimming, as well as in the Blacknose (*C. acronotus*) and the Silky (*C. falciformis*), this latter in captivity, when it was disturbed by divers doing maintenance work on the tanks. The Lemon Shark (*Negaprion brevirostris*) adopted the same posture when a diver approached it in gear that made him look like a Killer Whale; furthermore, it started to swim in tight circles forming figures-of-eight, and alternately opened and closed its mouth.

MOTIVATION FOR AN ATTACK

Accepting that sharks could not expect to eat a submersible, the point must be that they consider it either as a dangerous predator, or else as a competitor for food or another resource, or indeed as a combination of the two. The motivation of the Grey Reef Shark for that matter concerns other sharks as well, and it is therefore of interest to have as precise an idea of it as possible in order to understand the origins of attacks on humans by sharks in general.

Predation (attack to obtain food)

This motivation is not the most favourable one, for several obvious reasons:

- the submersible looks nothing like the usual food, which consists of small reef fish

- predators never warn their prey that they are about to attack

- in feeding periods (which can be provoked by baiting) the sharks are much less inclined to attack than in non-hunting periods; thus in one case a shark was buffeted while feeding on bait, and it merely left, the bait still in its mouth. In comparison, another shark, which had not been enticed with bait, was touched from behind by the submersible's dome against its tail: it immediately turned around and attacked on its side.

Earlier, in the 1970s, Balbridge and Williams had wondered whether attack by sharks was more combat than quest for food, observing notably that many attacks are of the "slashes" type, as if to inflict a wound rather than to remove flesh. They concluded that 50–75% of the listed cases of attack could have been caused not by hunger but by a form of aggressive behaviour directed towards victims, in an attitude of combat rather than of hunting.

In 1974, Barlow suggested that the surface wound could equally result from a conflict between wanting to bite completely and wanting to avoid a dangerous animal; the slash type attack has the advantage of very quickly protecting the aggressor from any counter-attack by the recipient, whether the latter be prey or rival.

Antipredation (attacks made to avoid being eaten)

There were a number of reasons in favour of attacks with preliminary display representing an antipredator behaviour:

- the submersible, besides being something new in the environment, looks and behaves more like a predator hunting, a big shark for instance, than like a prey or a competitor of the same species.

- the submersible's desire to corner or pursue increases the likelihood that it may be an aggressor.

- a number of sharks rapidly fled after having made their attack, as if the "animal" they had just attempted to intimidate might pursue them and even catch them again. This behaviour is comparable with that of the rat or the snake, which attack only when they are cornered and cannot escape.

The Grey Reef Shark's attack could also be "preventive", directed at an animal in the territory regarded as potentially dangerous, to adults but also to young. It has in fact been known since 1986 that the hollows of lagoons are the favoured breeding places of Grey Reef Sharks, and that in the same waters Tiger Sharks and Galapagos Sharks (*Carcharhinus galapagensis*) are known to devour other sharks, in particular the young ones. There would be nothing illogical in the submersible, but also any diver entering these regular breeding waters, being considered a potential aggressor and being intimidated as a preventive measure, or else driven off.

Competitive social motivation (attack to defend resources)

There are arguments both for and against competition as a motivation for display and attack, whether it involves competition in the form of territoriality (defending an area) or dominance (defending a social rank).

Territoriality has often been mentioned in the popular press, but it has never been possible to demonstrate this in any species of shark (Johnson, 1978; Barlow, 1974), and especially not in the Grey Reef Shark, which has never been seen to threaten or attack one of its own species. This last point is important, for territory defence is exercised more often within one and the same species than against another species (in the case of the submersible it could involve only "interspecific" and not "intraspecific" competition, as it is impossible for the apparatus to have been considered a congener by the shark). Competitors for the same food in any given habitat more usually belong to one species than to two different species.

Studies carried out in 1986 using telemetry showed, moreover, that Grey Reef Sharks travelled at least 15 kilometres daily, which would make a very large territory to "defend". In short, when an animal is defending its territory, it attacks until the latter is liberated. Whereas the Grey Reef Sharks which attacked the submersible did so only once, giving up the moment the craft no longer pestered them, even though it remained in the area.

What seems likely is that the shark defends a sort of "vital space", a sphere inside which any intruder is attacked or intimidated, a sphere which moves with the shark and is not connected with a particular territory. The reaction will be more or less acute depending on the type of intruder, on its activity and on that of the shark (whether, for example, it is searching for prey).

The difference in aggressiveness noted among the four species of sharks can corroborate previous hypotheses. The Grey Reef Shark was seen to be the only one to remain on the spot and to attack the submersible, whereas all the others slipped away. If we favour antipredation as the motivation, then habitat may be a sufficient explanation. The White-tip Reef Sharks stay near the bottom of the reef, ready to go back into the caves which can serve as a refuge against their predators. The Blackfins normally inhabit very shallow waters, such as the surface "platforms" of the coral, and could escape large predators by entering waters too shallow for them to be pursued there. The Silvertips live in deeper waters, unprotected, along the great ocean trenches, but doubtless their larger size makes them less vulnerable. The Grey Reef Sharks, smaller, and also inhabiting unprotected waters, are more vulnerable to predation, and so are obliged to have anti-predator reactions.

In conclusion, the aggravated behaviour of the Grey Reef Shark in event of threat is doubtless multifunctional with an antipredation and/or social competitive motivation. There is no doubt that the others have over the generations acquired the same motivations but without having integrated this kind of alarm bell that we call display swimming. If it were otherwise, sharks would no longer be quite so unpredictable, and they would therefore be less dangerous, and, who knows, perhaps less fascinating.

THE MARKINGS OF THE OCEANIC WHITE-TIP: LURE OR IDENTITY BADGE?

Body markings are often used in the animal world as means of communication. Would the same hold true for the shark? In certain cases these markings can be used for camouflage purposes (when irregular colouration is involved), but often this is not the case where particular distinctly pigmented areas such as the edges or the tips of the fins are concerned. Myrberg has put forward a particularly interesting hypothesis on the possible function of the white markings on the fins of the Oceanic White-tip Shark (C. *longimanus*). While conducting acoustic experiments on the oceanic sharks near the Bahamas in the 1970s, he often came across White-tip Sharks, which were attracted to the underwater microphones and remained in the vicinity, moving about slowly, almost lethargically. Their movements seemed to be made without any effort compared with those of other sharks he had observed over some years. These sharks were, however, capable of lightning-fast accelerations that propelled them out of sight at great speed. This happened after they had bitten the metal microphones or after the latter had suddenly emitted a loud noise. Such accelerations were able to explain a long-known fact about this species of shark; they very often took as prey some of the swiftest fish of the open sea: scombrids (tunny, mackerel, etc), various dolphinfishes, and even the white marlin. None the less the White-tip, even if it is very fast, is incapable of catching up such swift prey, or even of following them in open water. One theory had already been put forward to answer the mystery (Bullis, 1961): White-tips move about within groups of small surface-dwelling fish, at the time when the latter are being attacked by the fast-swimming fish. As these fast fish dash at their prey, they literally fling themselves into the open mouths of the sharks. Although this hypothesis is conceivable, it would involve a shark being positioned exactly in the line of a hunting fish, the latter's impetus being its downfall.

The hypothesis which resolves the problem is based on a visual effect that is noticed as soon as one encounters White-tips. As long as these sharks remain close, their outline is perfectly distinct and their presence unquestionable, but, once they move to the limit of visibility, hardly anything can be seen of them but their white-pigmented fins, the rest of the greyish body disappears. In such circumstances, nothing more than a few white patches can be seen moving slowly in formation. If there are two White-tips together, a group of white blobs is seen. Since the surrounding luminosity is poor, the contrast between the white and the grey bodies dissolving into the background increases. If we consider that a fish's vision can be compared to that of man, at least in the register of dark and light, then where man sees white patches, the fish's hunting instinct makes it see as many prey of reasonable colour and size. When it rushes at these prey and discovers its mistake, it is too late; the shark has already launched out to snap it up, and the fish's momentum prevents it from avoiding the trap. This scenario does seem to be the true one, and it does not exclude the possibility that the white fins might also be a signal used by the sharks to recognise each other.

This function as a lure as much as a recognition badge also accounts for the size of the first dorsal fin and the pectoral fins, which are so big they have been called paddles. One way of maximising the effectiveness of a lure is to increase its size so that it can be seen over a wider range.

From the point of view of hydrodynamics, the broad pectoral fins can only assist acceleration, and the big dorsal fin assist stabilisation when making a sudden movement forward.

It cannot be said which of the two benefits arising from the increase in fin size was retained by natural selection, but this is the kind of observation which, year after year, enables us to complete this or that piece of the puzzle, while awaiting the day when a picture of the whole will come to the fore.

ATTACKING NUCLEAR SUBMARINES

When modern submarines surface in shipping zones, the risk of collision with a ship would be great if certain marines such as the US Navy did not use a sounding probe which reaches the surface at the end of a cable while the submersible is still deep down. This floating probe is linked to the side of the submarine by a complex electric cable, covered in a rubber protective sheath 13 millimetres thick. In the 1980s, it was found that this covering was being regularly damaged, showing deep crescent-shaped lacerations, as if chiselled out.

At first it was thought that these were bites made by some animal, but as no biologist could identify any animal that might be responsible, a mechanical cause was sought. Exhaustive tests were undertaken but these were never able to reproduce the same effects. Finally, quite by chance, a scientist who was studying both whales and Great White Sharks was shown the laceration and was immediately able to provide the answer to the riddle. "Of course", he said, "all that is caused by the Cookie-cutter Shark, which feeds by removing whole mouthfuls of the skin of whales and large fish. The Cookie-cutter must be mistaking submarines for whales!" (See directory at end of book.)

Once the guilty party was identified, all that was needed was to cover the cables in very tough fibreglass, so that the Cookie-cutter would be wasting its time using its teeth on it.

Thirty or so submarines had been damaged in this way, not only the cables but also certain hull coverings had suffered. Fortunately no shark had been able to cut the cable completely; if it had, it is quite conceivable that a collision could have been caused by this 50 centimetre long shark that could have resulted in the loss of a submarine.

Many fishermen bring up nets that have been cut by the Cookie-cutter. The injuries they inflict on their hosts are characteristic and take the form of a hole 50 millimetres across from which the flesh has been torn as if with a stone-remover. Until recently, these wounds were attributed to bacterial infections or to some parasite or other.

THE SHARK'S COMPANIONS

To refer to them as "friends" would be too strong a term for these little fish which accompany the shark everywhere. Theirs is a purely selfish aim, with minimal benefit or none at all for their host. These fish are one of three species and are commensals, feeding on the scraps of prey shredded by the shark.

The first is best known by the name of pilot fish. It generally swims close to the shark's dorsal fin, only on rare occasions leaving on brief incursions into the immediate surroundings. Measuring 30

The pilot fish (*Naucrates ductor*): black and white stripes from head to tail

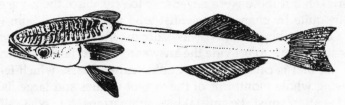

The remora (*Remora remora*): dark grey with no stripes

centimetres on average, its body is zebra-striped with between 5 and 7 vertical black bands. It should be made clear that it does not pilot the shark at all, being content instead to escort it. In fact it accompanies any large object, a boat as well as a shoal of tunnies, a turtle or a big manta ray. The shark, however, is a favourite host for several reasons. The food motivation is obvious, these little fish gorging on scraps too small for the shark. The protection offered by such a companion is also an important safeguard against predators; and finally, the shark's movement creates a wave above which it is easy to swim, like surfers on rollers.

The other commensal fish are the remoras, of which there are two: the actual remoras themselves, and the shark suckers, both attached firmly to the shark by discs which exert a very powerful suction. These discs are the modified vestiges of the first dorsal fin. If one day you get the chance to catch hold of such a fish, you will be able to pull it away only by a sideways or forwards movement, making it slide; pulling it upwards will only increase the suction power.

The remora is very widespread in all temperate seas, and may accompany its bearer into colder waters. Its average length is 60 centimetres.

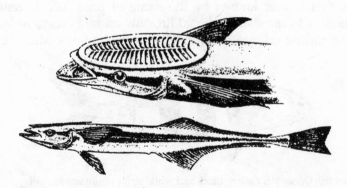

The shark sucker (*Echeneites naucrates*)

The shark sucker has a more slender body than the preceding species and is shorter, averaging 30 centimetres in length (the largest recorded measured 75 centimetres and weighed 17 kilos).

Sharks do not derive any advantage whatsoever from the pilot fish, and benefit slightly from the remoras insofar as the latter eat parasites in the form of small shrimps which are attached to their skin.

5

ATTACKS: INSTRUCTIVE DRAMAS

ACCURSED PLACES

AMANZIMTOTI IS A bathing resort twenty-seven kilometres south of Durban in South Africa. Its vast sandy beaches stretch out at the base of superb dunes covered with wild vegetation, and the scene as a whole is more likely to invite bucolic thoughts than macabre statistics.

One of these beaches, however, is the place where the most attacks in the world have taken place over the last fifty years: eleven since 1940, three of them fatal. Between 1974 and 1975 alone, four mishaps occurred within barely a year.

On 7th January 1974, at ten past two in the afternoon, Cornelius was bathing 100 metres from the shore in 2.5 metres of water in the midst of around five hundred swimmers. The sea was calm, the water temperature was 21.5°C and the visibility through the water about 1.5 metres. The River Umkomaas not far from there had brought down a lot of water in the previous few days, and a recent storm had caused several of the new nets set up against sharks just beyond the breakers to collapse. Cornelius was suddenly propelled above the surface and immediately realised that a shark had attacked him. He managed to punch it with his fist and immediately swam towards the shore while giving the alarm to all around him. Some surfers

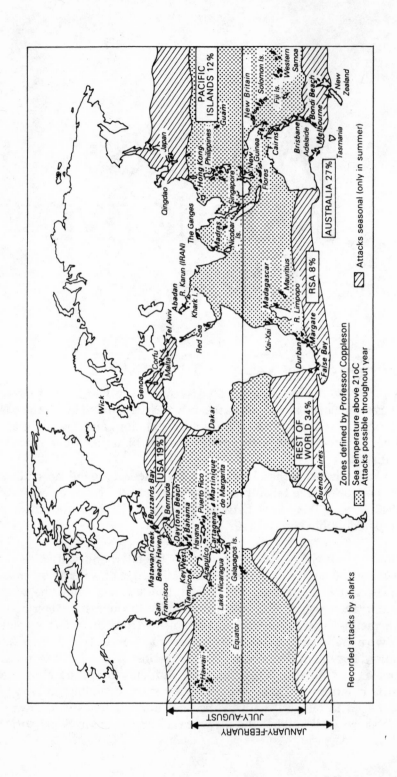

Recorded attacks by sharks

Zones defined by Professor Coppleson
Sea temperature above 21ºC
Attacks possible throughout year

Attacks seasonal (only in summer)

USA 19%
REST OF WORLD 34%
RSA 8%
AUSTRALIA 27%
PACIFIC ISLANDS 12%

JANUARY-FEBRUARY
JULY-AUGUST

San Francisco
Matawan Creek
Beach Haven
Buzzards Bay
Bermuda
Daytona Beach
Bahama
Key West
Havana
Puerto Rico
Tampico
Acapulco
Cartagena
Martinique
I. de Margarita
Lake Nicaragua
Galapagos Is.
Equator
Hawaii
Buenos Aires

Wick
Genoa
Corfu
Malta
Tel Aviv
Abadan
R. Karun (IRAN)
Khark I.
Red Sea
Dakar
Xai-Xai
Madagascar
Mauritius
R. Limpopo
Durban
Margate
False Bay

Qingdao
Japan
Hong Kong
Philippines
Guam
The Ganges
Madras
Singapore
Nicobar Is.
Flores
New Guinea
New Britain
Solomon Is.
Fiji Is.
Western Samoa
Cairns
Brisbane
Bondi Beach
Adelaide
Melbourne
Tasmania
New Zealand

108

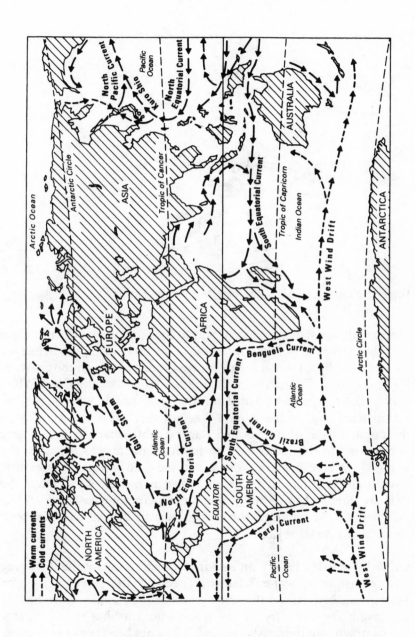

Diver or sportsman, no-one is safe from the curiosity of a shark.

came to his assistance and took him to hospital, where he recovered from his gashes within two months. An inquiry was immediately set up by the NSB, as it is each time an attack takes place on the coasts of Natal province. The nature of the injuries suggested the Great White Shark. Their asymmetry ruled out the Tiger Shark, whose jaws are symmetrical; and as the wound was not very deep, this eliminated the pointed teeth of the Spotted Ragged-tooth Shark, which is very widespread in the region. As no tooth fragment was found in the wound, the investigators were to confine themselves to conjecture. Regarding the choice of Cornelius rather than another swimmer, the data are thin: he had been wearing a red swimming costume, and he had had a cyst removed three days earlier; but he had not been wearing any metal objects.

On 13th February 1974, John Kendrick was training for the lifesaving championships in water 1.5 metres deep. In the preceding weeks, heavy rains had caused the rivers to break their banks, considerably increasing the turbidity of the water and washing much organic debris down into the coastal waters. It had not been possible to check the nets for seven days, and five of the nine were later found to have collapsed during the storm. The water temperature was 27°C, underwater visibility zero, and the sky sunny. For all these reasons bathing was forbidden. John Kendrick was swimming with his friend Joe Kool, who suddenly felt something rub against his side. He shouted to Kendrick: "Swim, quickly!" The latter later was to recount: "Joe was about five metres from me when he began yelling. As I turned around, a shark seized my leg. I heard a rumbling as the animal's powerful jaws started to shake me. Everything happened so fast that I hadn't realised what was going on. The shark released me after two seconds and I was pushed back into the breakers, which threw me up on to the sand. I succeeded in making my way backwards up the slope of the beach, holding my injured leg in front of me. It was only then that I realised – and my mind refused to accept what I saw – that where my calf should have been there were strips of flesh and muscle hanging like so many old rags, dripping with blood which poured out on to the sand." Kendrick had in fact been bitten three times, one bite fracturing his fibula, another passing through this bone near the knee, and the third taking off the fibula with the calf (see photo in sealed section). A witness applied finger pressure to the femoral artery and fitted a tourniquet, and this certainly saved Kendrick's life, his blood pressure having dropped to 74. The leg had to be amputated just below the knee.

The experts at the Natal Shark Board established the diameter of the arc of the bite (20.3 centimetres), which was confirmed by placing a jaw of the same size next to impressions taken of the wound. In the tibia, when examined closely, three small teeth marks were discovered, cutting deep into the bone, and enabling the probable perpetrator of the attack to be identified. The marks distinctly showed fine indentations on the edge of the teeth, which ruled out the Spotted Ragged-tooth as well as the Mako, whose teeth are pointed and smooth (see directory). A tooth fragment was examined under the microscope and this confirmed that the shark had to be one of the Carcharhiniformes. In this region, three could be implicated:

- *Carcharhinus amboinensis* (Pigeye Shark);

- *Carcharhinus leucas* (Bull Shark);

- *Carcharhinus obscurus* (Dusky Shark).

All three are common on the coasts of Natal and regularly attack swimmers or surfers. The length of the assailant was estimated at 1.9 metres, a very frequent size among sharks captured each day in the nets set by the NSB. Several factors had come together that day for an attack to take place:

- murky waters with nil visibility (recent rains)

- high temperature

- organic debris

- barrier of nets damaged (storm)

- channel created by unusual currents 3 metres out from the beach, in which the shark was lying in ambush

- noisy entry into the water by the two swimmers, as they later confirmed.

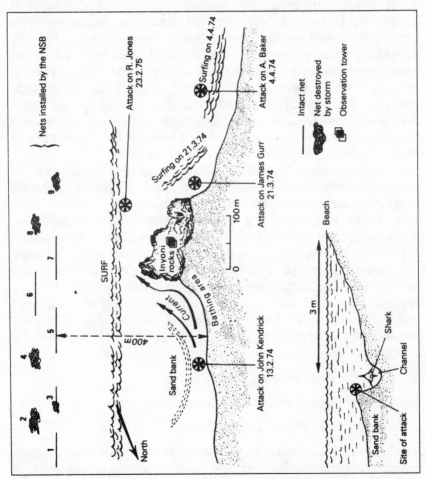

Plan of the beach at Amanzimtoti, on 13th February 1974, at the time when John Kendrick was attacked, and locations of the three other attacks some months later outside the protected zone

John Kendrick became famous, often appearing on television as one of the rare one-legged surfing experts.

On 21st March 1974, at four o'clock in the afternoon, 21 year old James Gurr was sitting on his surfboard when he caught sight of a shark fin charging straight at him. The underwater visibility was no more than 1.5 metres and James could not see the animal's contours. The water temperature was 24°C and bathing had been prohibited after four o'clock in the wake of the two previous attacks (sharks always attack more in the evening and at night). Gurr was dressed in a green and gold wetsuit and his ankle was held to the board by a small yellow cord. He was about 10 metres from the shore, directly above a channel 1.5 metres deep hollowed out of the sand by recent currents, just south of Inyoni rocks (see diagram); his feet were dangling in the water to help steady his board.

The impact of the shark knocked him off his board, and he attempted to get back as fast as he could. The board had been very heftily bitten. He then felt another impact against the side of his wetsuit. The shark's abrupt about-turn had surprised him but he managed to lie down on his belly on the plank, with legs up. "I felt totally powerless. The shark returned to the right and knocked me into the water. The board was 2 metres away, and I was preparing to pull myself back on again when a shove threw me on to my side. At the same time I felt the shark against my chest and beneath my arm. This really was the limit. Panic overwhelmed me and I started to paddle towards the shore. Fear increased my strength tenfold. Suddenly a terrific jolt found me face down in the water. Frantically, I got back on the board and continued paddling towards the shore, the shark zigzagging ahead of me. My terror increased further when a breaking wave propelled me over the top of the shark. I succeeded in reaching the breakers, which threw me up on to the sand."

Miraculously Gurr escaped without a scratch, and only the board had been bitten. The imprints on the upperside had a diameter of 19.8 centimetres and those on the underside 18.2 centimetres, corresponding to a shark of about 1.80 metres. (The shark's rapid about-turn between its first two attacks is in keeping with a relatively modest size.) Examination of the imprints showed teeth with a very sharp cutting edge. Witnesses having moreover clearly noted a uniform grey colour, it was concluded that the attacker was very probably a Dusky Shark (*Carcharhinus obscurus*). As in the previous case, we again find the existence of a channel near the shore, mediocre visibility, high temperature, bright colours, and a disregarding of the orders banning bathing.

On 4th April 1974, at half past four, Anthony Baker (aged 17) was surfing together with his brother 50 metres offshore, just south of Inyoni rocks. Here again, for the same reasons as before, bathing had been banned. The temperature was 24°C, visibility in the water less than 3 metres, and the sky sunny. A white sock covered Anthony's right foot, which was stuck into one of the aquaplane's straps. "I was facing the shore when I felt something pull at my foot, followed by a warm sensation. Because of the other attacks, I immediately knew that a shark had bitten me. I did not experience any pain, but I had heard that shark bites were painless. I turned to check if my two legs were still there. There was a lot of blood in the water, and the white sock had turned red. It was then that I took fright. I shouted a warning to Raymond, and started to paddle towards the shore. On the beach, I examined my foot and discovered an open gash in the heel, bleeding profusely. I thanked God that it wasn't more serious."

This wound was closed with nineteen suture clips, and its relatively modest size called to mind a small shark, without any doubt attracted by the sock covering the foot. A net with a 13 centimetre mesh was set up together with a line of hooks, which brought in eleven small-sized *Carcharhinus* sharks. There is nothing surprising about this when we know that sharks reach the proximity of the coast in the late afternoon, leaving again the following dawn for the open sea.

On 23rd February 1975, ten months later, at about 10.55 a.m., Russel Jones was surfing 100 metres off Inyoni rocks accompanied by two friends. Dressed in white boxer shorts, he was wearing around his right ankle a yellow sock attached to the surfboard. Jones had grazed his left knee half an hour earlier, but only superficially. He was seated on his board, his legs in the water, when he felt a bump against the toes of his right foot. Thinking it was a submerged rock, he pulled up his left leg, but it was then that he felt something really clamping his foot. "It was like the jaws of a steel trap on my leg." He shouted to give the alarm while continuing to pull at his leg, but the shark maintained its grip and its snout broke the surface a rounded snout, grey in colour. The shark dived, dragging the leg towards the bottom and Jones off his board. Jones panicked at the thought that the shark was now going to set about his body. He pulled with all his might and felt his foot come away from the leg. He got back on the board, pulled his leg out of the water, and was so shattered by what he saw that he felt himself fainting. He felt no pain until when he tried to walk on his stump on the sand. The tibia and the fibula were sticking out of the torn flesh and serrated teeth marks were found on them. The foot had been bitten off and the flesh torn away, as if by a stone-remover, 12 centimetres above the ankle, on the side which had been wearing the yellow sock as in the preceding case. Examination of the wounds and of the marks on the bones led to the attacker being identified as a *Carcharhinus* species, probably *obscurus*, about 1.5 metres in size.

This macabre series of events was not the work of a single solitary shark, as was seen for example in New Jersey in 1916, but occurred in a spot particularly favourable for sharks to reside:

- presence of a major tidal bore or of an underwater channel allowing ambush, and of rocks allowing a varied habitat

- nearby rivers

- sea beds rich in animal life

These accidents paradoxically confirmed the effectiveness of the nets laid by the Natal Shark Board based at Durban, since the mishaps took place when the nets were not all in position or when the victims did not respect the ban.

SHIPWRECKS: SITES OF UNSUSPECTED CARNAGE

Shipwrecks and air disasters are very often the occasion of out-and-out slaughter which is not always accounted for, because the victims are no longer there to give evidence and because the direct witnesses have only a partial view of the events that are going on around them.

On 26th February 1852, the English frigate *Birkenhead* struck a reef south of the Cape of Good Hope at 2 o'clock in the morning, only 1600 metres off Danger Point. On board were 490 English soldiers, with 25 of their wives and 31 children, in addition to a crew of 134 men under the command of Captain Robert Salmond.

In the first few minutes following the collision, confusion reigned on board as the soldiers, the sailors and the passengers all rushed on to deck to escape the flood of water invading the front of the ship. The officers quickly re-established composure by ordering the men to regroup at the aft end of the boat, and Colonel Seton commanded his officers to see to it that all of Captain Salmond's orders were carried out immediately.

The latter ordered the women and children to board a rowboat, detailing a sub-lieutenant and a sergeant to separate the women from their husbands by force. A second boat was launched with 30 men on board. There were no more lifeboats for the 600 men left aboard the *Birkenhead*, whose deck was leaning more and more towards the bow. Many passengers trapped below decks had already drowned, and others had died when the masts crashed down on to the deck. Others still had been hurled overboard, and screams of terror were already coming from the surrounding waters. The victims could not see the dramas unfolding around them in the night, but they guessed that the screams were those of swimmers snatched towards the depths by sharks. There were 200 men left on deck when Captain Salmond climbed a few metres up the foremast to shout for the men's attention: "Every man for himself now. Your only chance if you can swim is to jump into the water and try to cling to anything floating, but I beg you to avoid the rowboat with the women and children, it's already overloaded. I am asking you in fact to stay where you are."

Three men only were to jump overboard, and, of the rest of the 200, nobody moved, stoically obeying orders. A few moments after Salmond's harangue, the ship's hull broke in two, all the deck having disappeared under the water while the stern rose above the surface to completely expose the rudder. A surviving officer was later to write: "Every man did what he had been ordered to do, and there was not a shout, not a murmur among them, until the ship finally went down... The officers had received their orders and had executed them, as if the men were embarking instead of going straight to the bottom of the sea. There was only one difference: I have never seen embarkation carried out with so little noise or confusion."

The *Birkenhead* sank exactly thirty minutes after hitting the reef. The sea was red with blood, and on the surface could be seen what appeared to be hardly recognisable human remains, torn up by sharks. Lieutenant Girardot described to his father: "I remained on deck until the boat sank. I was dragged under the water by the suction and a man caught my leg. I succeeded in freeing myself by kicking him and got to the surface, where I hung on to some pieces of wood. I remained in the water for five hours... The sea was so high that many perished trying to reach land. Practically all those who found themselves in the water without their clothes were taken by sharks; hundreds of them surrounded us, and I saw a number of men seized right next to me, but, as I was dressed... they preferred the others."

More than 60 men managed to swim the 1600 metres separating them from the coast, but the majority of the *Birkenhead's* passengers could not swim, including Colonel Seton, who drowned. Captain Salmond was hurled overboard and killed by the falling masts. The shipwreck claimed 455 lives, and the proportion of deaths due to the sharks was certainly of the same order of magnitude as that due to the ship and to the sea. When the account of the drama reached London on 15th April 1852, the complete list of victims was published, but not a word was mentioned about sharks.

On 18th November 1942, at nine fifteen in the morning, in the same region but on the Indian Ocean side, the English steamer *Nova Scotia* was sunk by a German U-boat 50 kilometres off St Lucia in Natal province. As well as the crew, the boat was carrying 765 Italian prisoners of war and 134 South African soldiers returning from the Middle East to Durban. Most of the lifeboats were on fire, and hundreds of survivors found themselves in the sea in life jackets or clinging to ordinary wooden or rubber life rafts. George Kennaugh of Johannesburg testified as follows: "Suddenly there were two terrible explosions; we had just been torpedoed. I tried to reach my lifebelt but the boat was already listing and I slipped on the deck, which was oily. I found myself in the sea, wearing nothing but my swimsuit. I swam through the black oil-covered water and clung on to a floating wooden oar. There were about a hundred men swimming around me, clinging to bits of wreckage or rafts. Another soldier from my regiment clutched the same spar as myself. He was wearing a life jacket. We drifted all night, carried by a strong current. At dawn there was no longer any oil but still many survivors around us. As soon as it got light my companion told me that he would rather die than remain thus, clinging hopelessly to a piece of wood. He told me that he was going to let go and refused to listen to me when I tried to dissuade him. So I asked him to let me have his life jacket before letting himself drift. As he was loosening the straps, he suddenly screamed and the entire top half of his body was literally lifted out of the water. When he fell back again, the sea was covered with blood and I saw that his foot had been severed. At this point I caught a glimpse of the grey form of a shark swimming excitedly around us, and I moved away as fast as I could. Then several sharks gathered around me, about 2 to 2.5 metres away. Now and again one of them would head straight for me, and I struck the water with all my might to make it change its mind, which seemed to work. I caught sight of a life raft with some Italians and a South African sergeant. I managed to climb aboard and we were able to survive thanks to a chest which contained some water and food. The sharks continued to circle us and we struck them with blows of the oars to keep them away. Sixty-seven hours after the ship had been torpedoed we were rescued by a Portuguese sloop."

The Portuguese sailors had to strike the sharks with grappling irons to keep them at a distance during the rescue operations. A total of 192 survivors was rescued, and of the 850 dead, it can be estimated that more than half were killed by sharks. All these men were in fact in the full prime of life, they could swim, and the sea temperature at that time in these waters would have allowed them to wait for help.

Still in the same waters, the *City of Cairo* was sunk on 2nd October five days out of Cape Town. Captain Angus MacDonald was on board a longboat with 54 survivors: "Before midnight on the very first day we saw our first sharks. They were enormous, and as they slipped beneath the boat fore and aft it seemed that they would bump into us and cause us to capsize. They were content to brush against the boat each time, and at no moment did they leave us... Several comrades died during the night, and we had to throw them in the sea at first light. That morning the sharks appeared in large numbers, and the water literally seethed the moment the bodies touched the surface. On the fifteenth day, one of the sailors decided to commit suicide by drowning himself, and threw himself over the stern. He had forgotten to remove his life jacket, and, as we were too weak to come to his aid, it was the sharks that killed him, not drowning."

The survivors were picked up by a German ship, the Rhakotis, but this was in turn sunk by a British cruiser. They finally found themselves aboard a U-boat.

These stories underline the way sharks automatically appear as soon as there is a shipwreck in certain tropical waters, and seafarers as well as builders of lifeboats should be aware of this rule. This would allow the former to equip themselves appropriately, and would stop the latter building any old thing (the sole French manufacturer persists in making boats with yellow bottoms, a colour particularly attractive to sharks).

This omnipresence of sharks in tropical waters is also very prejudicial to the survival of shipwreck victims for another reason, apart from the risk of physical attack. The psychological impact on physically and mentally weakened survivors is obvious, but sharks are also an impediment to a very important survival measure: bathing. In an earlier book, on surviving shipwrecks at sea (*Naufragés comment survivre en mer*, Filippachi, 1989), I had a chance to develop the multiple benefits of baths for shipwreck victims: better heat regulation, reduction in dehydration, disappearance of "pathological ankylosis" (stiffness), removal of salt from the skin, reduction in dermatoses, and psychological benefit, etc.

Just a few days before the English surrendered to the Japanese at Singapore in 1942, three soldiers escaped to Sumatra, where they procured a tiny boat 5 metres by 1.4 metres. With this dinghy which barely had room for three to sit, they undertook to sail to Australia, 2600 kilometres away! They suffered 125 days of nightmares at sea, to find themselves finally back in Sumatra, 190 kilometres from their point of departure. It was not only their inexperience in the matter of navigation that made the expedition futile, but also the presence everywhere of sharks that drove them almost mad. The heat was stifling on board their little open boat, and the feeling of oppressiveness difficult to bear on a windless sea. They were constantly obsessed with the thought of diving overboard to refresh themselves in the blue water which seemed to invite them in continuously. But the sharks were always there, day and night, non-stop. Day after day their shadows were like evil guardian angels following the boat at 2 or 3 metres. Not once did they dare to bathe. One day when the three men made a small fire to prepare a meal, one of the sharks struck the boat with such force that it destroyed the rudder.

In January 1942, a US Navy torpedo plane, caught in a storm, ditched in the mid Pacific. The three crewmen took refuge in the dinghy, which measured 2.5 by 1.25 metres, together with their remaining equipment: a pocket knife, a pistol and a pair of pliers. By the fifth day the lack of space bothered them considerably, but sharks prevented them from taking exercise by bathing, leaving them to roast in the tropical furnace. Among the three men, the radio-operator Aldrick demonstrated an exceptional skill in catching fish, which he skewered with his knife.

It was thanks to him that the three men were to survive thirty-four days. One night, Aldrick plunged his hand into the water to check the direction of the current. Unfortunately, a shark was on the lookout and immediately seized his fingers, cutting them to the bone. The wounds very quickly turned septic, but one of his companions lanced them, allowing the pus to escape and Aldrick to resume his miraculous fishing. This shark was small in size, but a few days later they passed through a zone overrun with Leopard Sharks "so aggressive that they threatened at any moment to overturn our dinghy. At one stage we were obliged to beat off one of the sharks by hammering on its snout with our fists, and we succeeded in killing another of them with a pistol shot, before rust made our only firearm unusable."

We can say that all of the stories recounting long periods of drifting in tropical waters mention the virtually constant presence of sharks. The small ones are always in the immediate vicinity, while the big ones are most often seen at nightfall and at daybreak.

It is clear that the perfecting of an effective shark repellent will not only allow the man in the sea to escape sharks, but also allow the shipwreck victim, "hemmed in" in the dinghy, to enter the water regularly without getting attacked.

Some more determined people have successfully taken the risk; for example, a handful of survivors from the British cruiser *Avocet*, who found themselves on a makeshift raft built of empty oil drums. The executive officer had taken charge of the small craft and kept a strict and detailed log:

"During the afternoon calm, when the heat is intense, the men asked if they could bathe. At first I refused for fear of the sharks that surrounded us, and then eventually I decided that it was better to take the risk rather than see the men become mad under the influence of thirst. Bathing was therefore organised in such a way that there is never more than three men in the water at the same time, the others keeping an eye on the sharks that are always nearby. Every day, each of the men spends a moment in the water, with a rope around his waist. As soon as the men catch sight of a grey shadow rapidly approaching, they abruptly pull their companions out of the water. We are always anxious lest one of the sharks should catch us."

Some shipwreck victims have experienced even greater anguish, like the three men who found themselves in a tiny dinghy 1.5 metres square. With the excess load, the dinghy floated with 8 centimetres below the surface and the sharks continually brushed them, gliding just underneath and gently nudging them. The three wretched men remained sitting in the half-submerged dinghy for 12 days. Twelve days during which time they prayed for help and suffered the most discomfort imaginable. One of the victims succumbed; the other two were rescued.

One morning in November 1918, the steamer *Una* sank on a shallow bank 100 kilometres north of Santo Domingo in the Caribbean Sea. The officers and crew boarded the lifeboats, but there were not enough on board and 75 workers had to board very rudimentary life rafts. The sea was calm, and in less than an hour the waters were invaded by sharks varying in size from 2.5 metres to 5 metres.

Some rafts were overloaded, and as soon as a man fell in the water he was immediately torn to pieces by the sharks. As time passed, the sharks became more and more aggressive and even attacked the terrified men on the rafts, trying to make them fall off with a lash of their tails. Others put their heads on to the raft and seized the wretched survivors by a hand, an arm or a leg, immediately dragging them into the water. Some rafts tilted so much under the sharks' weight that the exhausted victims could no longer stand and slid into the water, where they were dismembered within a few seconds. A number of sharks tried to lift up the rafts with their backs and in this way managed to throw a few additional victims off balance. The men had nothing to defend themselves with except a few oars, which were rapidly sheared or torn from their hands.

The nightmare was to last for several days, and the thought of having escaped a sea disaster only to find themselves dismembered by the hideous teeth of the sharks was to drive several of the unfortunates to madness. Many threw themselves in the sea, directly into the jaws of their persecutors. Of the 75 men, only a handful that had been spared by the sharks were rescued.

Nobody will ever know how many hundreds of rafts or dinghies and their occupants have disappeared in this way, harassed by sharks. Considering that each year there are about 50,000 shipwreck castaways in the world, a good half of these in the tropics, I should think that several hundred of them may become victims of sharks. For those who would accuse me of pessimism or sensationalism (even though, on the contrary, I pay particular attention in my books to dissecting the risks objectively, and especially to giving recipes for surviving any extreme conditions), I shall limit myself to recalling the example of the *Dona Paz*. This ship sank in December 1987 in the Sea of China, between the Philippine Islands. Built in 1963 to shuttle back and forth between two ports in south Japan, the ferry was authorised to carry 608 passengers. Hastily reconditioned for the Philippine company Sulpicio, the official capacity was raised to 1500 passengers. As the end of year celebrations approach, the Filipinos have a custom of moving from island to island. Being poor, they had no hesitation in huddling together in twos, threes, even fours per berth. At Manila, it is thought that the population that boarded the *Dona Paz* on that day of 20 December 1987 must have been between 3000 and 4000 people. I shall not bring up again the human causes of this second largest shipwreck of all time, but linger only on all the factors which were inevitably bound to attract all the sharks in the area.

It was ten o'clock at night when the small oil tanker *Victor* struck the *Dona Paz* on her port side, just behind the forecastle at a height level with the passenger decks. The *Victor* was carrying 880 barrels of kerosene which immediately caught fire. The fire spread to the ferry, and explosions followed one after the other. The blazing fuel spread over the sea and around the two ships. The panic among the passengers was dreadful and the crew were completely unable to cope. All those who succeeded in getting over the handrail and into the sea were burnt instantaneously. Many others were blown to shreds by the explosions. No lifeboat could be launched, and the only survivors would be those who had sufficient breath to swim underwater for the time necessary to emerge on the other side of the flames. This is why none of the children or old people survived. In fact the number of survivors was absurdly low: only 2 women and 23 men.

Fewer than 1% survivors, if we accept that there were between 3000 and 4000 passengers. The reasons for these staggering consequences were many, and the part played by sharks in completing the carnage was certainly considerable. Let us first stress the presence of numerous large predators in the China Sea: Great White, Tiger Shark, Sand Tiger Shark, Oceanic White-tip, Blue Shark, etc (see maps in the directory), and the preference of many of them for nocturnal hunting activity. Observe moreover the multitude of "provocative factors" surrounding the shipwreck: the noise (dull explosions, metallic crashing sounds carrying a very long way in the water, passengers' screams), the smell of the blood and the burnt flesh of thousands of victims, the innumerable vibrations and movements easily detectable by the sharks' specialised organs, the bright contrasts illuminating the surface of the water like daylight, the variations in salinity and in electric field induced by the liquids and other objects.

The next day, the helicopters and the many boats sent to the scene did not find a single body among the patches of fuel and the thousands of floating objects, proof that the sharks had been there before them. Only 300 mutilated corpses were to be fished out of the water far from the site of the shipwreck, and for some weeks the Philippine fishermen found human remains in the stomachs of many of the sharks they caught. They even went so far as to give up one of their national dishes, the "lapu-lapu", based on grouper.

The time sharks take to appear at the scene of an air or sea disaster in tropical or subtropical waters is generally very short, varying from 30 minutes to 24 hours. Apart from the sharks which happen to be in the immediate vicinity of the accident, in general they allow the victims of shipwrecks time to board the lifeboats. According to the testimonies of the two and a half thousand pilots or shipwreck victims in the Second World War who found themselves in such circumstances, sharks sometimes appeared within half an hour, but most often after an interval of 24 hours at most. As the days passed, they became more and more familiar, leaping out of the water and splashing the occupants of the lifeboats, striking the fragile edges with their enormous tails, lifting the boat and those in it several tens of centimetres, biting or breaking the paddles, swallowing the fish sheltering beneath the boat as well as any hands and feet mistakenly dangled in the water.

Whatever the size of the lifeboat, it needs only a few days for microalgae and minute crustaceans to attach themselves to its surface, very quickly attracting small fish which will be eaten by larger ones, themselves devoured by the big predators. The small shadowy cloud that the silhouette of a lifeboat represents in the bright vastness of the surface of the ocean is a point of attraction for the curiosity of all the animals in the food chain, from the smallest shrimp up to the biggest of the sharks. When in addition this anonymous shadow releases at regular intervals organic matter stimulating the sharks' acute sense of smell, the latter waste no time in drawing near the boat and may stay for days and days, even for whole weeks on end. It is obvious that all lifeboats ought to contain the equivalent of "dustbin bags" in which shipwreck victims could accumulate all their organic debris, which they could thus dispose of regularly without leaving any trace of odour. When we look at the dimensions, the weight and the absurd price of a few dozen of these plastic sacks, it is a wonder why they are not an integral part of the regular equipment of any lifeboat worthy of that name.

If someone should die in a lifeboat, it is imperative that he be given up to the sea during the night following his demise to avoid attracting sharks. Two shipwreck victims drifted for nine days on a raft in the Gulf of Mexico, and one of them died of dehydration. Only a few hours after his death, his companion was horrified to see the sharks, which up to then had not appeared aggressive, shoot half out of the water and seize the body to feast on it. The thought of taking part in the cutting up of a friend's body has often led shipwreck victims to keep the corpse for too long, thus attracting a larger number of sharks around them. The anonymity of the night will avoid the distaste of such a spectacle.

The worst situation imaginable for a shipwreck victim is being thrown into the water inadequately protected, that is to say hardly clothed, and barely supported by a life jacket. Survival in these conditions depends first on the temperature of the water: 15 to 30 minutes in water at 0°C if one is clothed, 1 hour 30 minutes at most in water at 10°C, and over 24 hours in water at 25°C. At 25°C, we are talking of tropical waters, and then survival depends on how rapidly the sharks arrive, while at the same time recognising that many victims have managed to be saved even when they were surrounded.

When, in 1950, a transport aircraft linking Puerto Rico and Miami ditched in the sea off Florida, many passengers survived, floating in their life jackets. Shortly afterwards, the pilot of a plane that flew over them signalled that a number of them were being attacked by a horde of sharks. Each tragedy was made obvious to him by the red patches that appeared on the orange specks of the lifejackets. Another spotter-plane pilot was likewise traumatised by the spectacle that presented itself in 1987 in the Caribbean. At the beginning of October that year, in a clandestine operation, 168 Dominicans left their country for the United States, via Puerto Rico. Aboard the boat were women, children, old people and men packed in together. For some unknown reason the boat capsized. All the passengers found themselves in the water in a warm tropical sea, which should allow for relatively long survival. The castaways being only eight kilometres from the coast, they undertook to swim towards it. Their efforts were to last 12 hours, in the course of which they had to face the blazing sun, a current running against them, and, above all, uncompromising aggressors, the sharks attracted by their injuries.

Head of civil defence, Eugenio Cabral, flew over the spot in a helicopter and looked on, powerless, at a horrifying spectacle. "The sharks seemed to have gone mad", he explained. "At first they were devouring only the corpses, but very quickly they attacked all over the place. They were big sharks, makos and hammerheads, and the water had turned red." Of the 168 passengers, only eight survivors were rescued!

It is obvious that these poor wretches were victims of that aspect of shark behaviour which is not regular but is incredibly dangerous: the feeding frenzy (the way to survive this is described elsewhere in this book). Knowing the uncontrollable murderous excitation that seizes sharks at such times, and the terrible vulnerability of a man in the sea, it is easy to understand how only a minute minority escaped the massacre.

Still on the subject of wrecks in tropical waters, Christian Troebst reports the case of a relatively lucky victim.

"During the war, an American pilot fell into the sea with two other airmen just off the South American coast. At the end of five hours one of the latter died of exhaustion, and the pilot started to swim pushing the body in front of him. Suddenly, something seized and shook the body, which then disappeared for good beneath the water. The survivors continued to swim through the night, but, after a few hours, the second airman also died. The pilot again started to push him ahead of him. Meanwhile the moon had risen, and the brightness suddenly enabled him to distinguish the dorsal fins of a large number of sharks swimming in a circle around him. Once more a jolt shook the corpse, which briefly sank under the water and then rose again to the surface, now without feet. Horrified, the swimmer turned it and grabbed it by the shoulders. The body immediately submerged a second time, only to reappear and resubmerge. The sharks devoured it little by little up to the shoulders. At dawn they began to attack the pilot. The latter knew that he was very close to the shore, and yelling and beating the water frenetically, he made it to land unharmed."

An American sailor whose destroyer was sunk off Guadalcanal reported: "I had been drifting for eleven hours when I suddenly felt my left foot itching. I lifted it above the water: it was dripping with blood. I immersed my head and I saw the shark charging at me. I shook my arms and legs violently and it passed very close, brushing against me. It turned about on itself and came back straight at me. I clenched my fist and delivered it a blow on the jaw, with all my might. It moved away not without having torn off a piece of my left hand; it attacked again and, once more, I hammered it in the eyes and the nose. When it moved off, I discovered that it had slashed my left arm. My heel had disappeared, too. At this point a lifeboat approached. I frantically beckoned it and forgot the shark. It tore away a piece of my hip, exposing the bone. Then I was hoisted into the boat."

Just before Christmas 1948, Tony Latona, a 13 year-old boy, was recovered on a beach in Cuba. He was in a critical condition and around his waist was wearing a lifebuoy in a sorry state. He had just spent forty hours in the water, and his story was difficult to believe at first. He told how he had been playing with another boy, 14 year-old Bent Jeppsen, on the after deck of the Danish ship *Grete Maersk* when Jeppsen went overboard about 14 kilometres from Cape

Maisi in Cuba. Tony threw Jeppsen a lifebuoy, then jumped overboard himself to help him. Their shouts of course were not heard and the ship disappeared. They had been in the water for two hours when they saw sharks arriving to attack them. One of them attacked Jeppsen and left two deep gashes in his left foot. Latona's story went on: "We banged and banged until the sharks moved away. I told Jeppsen that the blood in the water would send the sharks crazy. I told him to take off his trousers and tie them around his foot to help stop the bleeding. We did not see the sharks any more, but they can't have been very far away because, an hour later, when Jeppsen's trousers went, the sharks were back upon us within a few minutes. They passed just behind me and tried to grab Jeppsen. We continued to keep them at a distance, but they came back every quarter of an hour. And then a shark caught him again by the same foot. He complained about pain. The sharks came back more often, taking less and less notice of our efforts to keep them off. Soon enough another one bit Jeppsen under the arm. He wept when the shark tore away his flesh. Another one arrived which snatched away his knee. He howled and began to slip down. He sank beneath the water screaming "My foot!". He emerged again, screaming and fighting, and then he vanished again. That was the last time I saw him. I saw some blood in the water, so I sat in the lifebuoy and I kept my feet on the edges, above the water, paddling with my hands until I was too tired. When day broke I was near the coast, but the daytime currents pushed me back out to sea. The following night the sharks returned, one of them removing the bottom of my trousers. On the morning of the second day, a current finally brought me ashore." Again we notice in this example the very selective attraction of the shark to the prey that is bleeding, maintaining a total indifference to any other potential prey, even very close by. This behaviour is very different from the feeding frenzy, in the course of which any object inert or living is indiscriminately torn up and swallowed.

Commander Kabat also found out to his cost how relentlessly a shark can pursue its prey when his destroyer, the *Duncan*, sank off Guadalcanal in 1942 and he found himself in the water for a whole night with an old kapok life jacket and two small empty powder kegs as his only means of keeping afloat. Shortly after dusk he felt an itching in the region of his left foot and discovered that it was bleeding. He then noticed the brownish shadow of a shark less than three metres from him. The shark swam around him several times and then attacked again. Kabat tried to hold it off by punching it. After the animal had left, he found that a piece of flesh had been removed from his left hand. At intervals of about fifteen minutes he was attacked and wounded – a little more each time. At first his big toe was removed, then a piece from his right hip, then another from his left shoulder, from his right hand, from his buttocks. Kabat noted: "When it was not planting its teeth in my flesh, its rasping hide was removing great lumps of skin." In the course of the attacks which followed, his thigh was sliced into so deeply that the femur was visible.

Statistical study of attacks on people in the sea tends to prove that the risk is greater if the victim is not clothed. During the Second World War, the engine of a reconnaissance plane broke down 110 kilometres east of the Wallis Islands and 400 kilometres west of Western Samoa in the open Pacific. Lieutenant Reading managed to put down his craft without too much damage but he was knocked senseless by the impact. It was his radio operator Ahmond who managed to get him out of the cockpit and to release his life jacket before the plane sank. On contact with the water it was not long before Reading regained consciousness, and the two men lost in the open ocean opened their fluorescein bags in the hope of being picked up before night came. They tethered themselves to each other using the strings of their life jackets, and they waited. Reading was clothed and Ahmond was wearing shorts.

After about half an hour the sharks made their appearance. Shortly afterwards, Ahmond signalled that something had knocked against his right foot. His foot was bleeding and he tried to keep it out of the water. The sharks came back and the two men were dragged beneath the surface for a moment. The water around them was reddened by the blood and there were now five sharks harassing them. Not only did Ahmond's right leg bear several wounds, but his left thigh was deeply gashed as well. He felt no pain even though the sharks continued to harass him. Reading struck those which passed within his reach with his binoculars, but almost at once the beasts returned to attack Ahmond. The two men disappeared underwater again, and, when they resurfaced, Reading discovered that he was separated from his comrade. It was then that he was half knocked out by a flick of a tail on his chin, but the shark responsible for the blow was still only interested in Ahmond, who was now completely underwater and neither he nor his life jacket emerged again. The monsters continued circling around and from time to time Reading could feel them brushing his feet, but he was still not attacked. He was rescued after sixteen hours in the water, sixteen hours of terrible anguish surrounded by sharks and by the remains of the man who had saved his life.

It is not only being clothed that can reduce the risk of being attacked. When there are several people in the water it is also essential to group together back to back. Being in a coherent group enables heat loss to be reduced appreciably and, in tropical waters, a look out can be kept for sharks, this also permitting one to shield oneself against attack without worrying about what is going on behind one's back. It is also possible that the combined bulk of several people in the water joined together in a "pack" will, by its very nature, arouse the shark's distrust as much as its curiosity.

Dr Llano reports that the longest period of survival of men in the sea in shark-infested waters involved a group of soldiers who spent 42 hours shoulder to shoulder in the water. Of the twelve who survived without being attacked, eight were fully clothed, even though some of them removed or lost their shoes.

In 1942, the *Dorsetshire* was sunk by mines in the middle of the Indian Ocean. Commander Agar very quickly realised that, for the hundreds of men around him in the sea, the danger came above all from sharks. He ordered his men to gather together all the dead bodies floating around them, so all the survivors positioned themselves backs facing inwards around the macabre "platform". They remained like this in the open sea for 36 hours, battling against the sharks which gathered around the 60 corpses. The sharks were able to divert their aggression on to the dead men, sparing the survivors who seemed to be less easy prey.

Vic Beaver was a 74 year old Australian who held a number of national game-fishing records. On 11th March 1977, he set off aboard his boat in Brisbane Bay accompanied by two friends who liked to share his favourite pastime. It was well into the night when one of these, Harrison, discerned through the rain the lights of a cargo vessel on a collision course towards them. When the 2500 tonne ship rammed their boat, they sank immediately and found themselves in the sea, their only means of buoyancy a one metre ice container and an air bed. The three men were to cling to the container as best they could for 36 hours. Small sharks came and threatened them, and then a big one joined them. When this shark attacked Vic Beaver, Harrison tried hard to hold on to Vic and to deter his aggressor by kicking and punching it, but to no avail. Later he recounted: "I tried to keep Vic with me but he was being pulled out of the container. Vic simply said to me: "It's got me again. So long, friends, that's how it goes."

And then he vanished. That's all he said when the shark took him. John and I tried to huddle together inside the ice box, but we could only protect our heads and our shoulders and the shark was still beneath us. I punched and kicked it and cut myself slightly. The shark then began to circle us again and John cried out: "It's had my foot!". He told me to stay in the box, and then the shark dragged him under the water. I tried to climb up the box for safety, but the bastard tried to climb with me."

An hour after this second fatal attack, Harrison was rescued by the crew of another fishing boat. The shark must have been either a Great White or a Tiger, as both species live in the bay.

Coppleson reports another drama that occurred unexpectedly, again in Australia, in North Queensland. Ray Boundy was skipper of a 14 metre fishing boat trawling near the Townsville reefs. One of the trawl's hoists happened to break, throwing it off balance, and then a huge wave overturned the boat. Boundy took refuge on the keel along with his crewmate Denis Murphy, aged 24, and Linda Horton, aged 21. They decided to leave the sinking wreck, taking with them a surfboard, a lifebuoy and some pieces of polystyrene, and to try and reach the nearby reefs where they could be spotted.

At dawn on 25th July 1983 they were no more than 8 kilometres from the town of Lodestone, but nobody saw them, not even the aircraft that passed above them in its searches. Shortly after nightfall, a shark appeared and began to push at the board, the pieces of foam, the lifebuoy and the three castaways. "We didn't take too much notice of this", Boundy recounted, "thinking that if we didn't annoy it, it would leave us alone." The shark at first took an interest in Boundy's leg, but he kicked it with his other foot and the animal disappeared.

Ten minutes later, a big wave knocked the three castaways into the sea and the shark immediately returned. Murphy began to yell: "It's had my leg, the bastard's had my leg!" and then, a few seconds later: "This time, that's it; you and Lindy, go, get away", and he swam three or four breast strokes towards the shark. In the darkness Boundy and Lindy heard curses together with pounding in the water testifying to a desperate struggle between man and animal. They saw their companion's body emerge face down, and then disappear into the shark's mouth.

Everything went quiet for the next two hours, and then the monster came back to circle them from 4 o'clock in the morning onwards. Boundy continues: "Lindy was sitting in the snaffle of the lifebuoy with her feet out of the water resting on a sheet of foam. I was practically sure that it was the same shark that came back. This time it approached slowly, then suddenly seized Lindy, its enormous jaws around her arms and her chest while she was still seated in the buoy, shaking her three or four times. She let out only a small cry just as the shark crushed her rib cage, and I knew almost instantaneously that she was dead."

Boundy used two pieces of foam by way of paddles in an attempt to get away, but just after sunrise the shark reappeared again, to circle around him: "I thought I would never escape, for it was circling nearer and nearer to me, and then I caught sight of a reef protruding above the surface." Boundy succeeded in surfing as far as the reef by making use of a piece of foam. There a plane spotted him and a helicopter of the RAAF picked him up. The attacker or attackers were probably Tiger Sharks or large Whaler Sharks (*Carcharhinus obscurus*), very widespread in this zone (see the directory).

To conclude this discussion of relations between shipwreck and shark attack, let us note that there is no survivor who has spent more than 24 hours in a tropical sea, who does not mention the menacing presence of sharks, permanent or temporary, even though that menace has not always ended in an attack. The curiosity with which sharks regard an inflatable dinghy is the same as that for any inert object which suddenly installs itself on the homogeneous vastness of the surface of the waters. Although the dinghy generally has a régular shape, without floating appendages and moving parts, and though it has no organic odour, its colour is uniform and it does not emit any noise, what attracts the sharks is the fish that accompany the dinghy, themselves arrested by the shadow of the boat, and the biotope that establishes itself on its surface after a few days. If, moreover, accident victims do throw fish scraps, blood and waste matter over the side, the sharks will very quickly become permanent companions.

If the shipwreck victim is yet more unfortunate, and finds himself not in a life-raft, but in the sea – sharks will be more attracted to him if he splashes about, emits vibrations and odours and if he is unclothed.

BOATS ARE NOT ALWAYS SAFE FROM ATTACK

People with fertile imaginations or those who have seen the film *Jaws* may find themselves wondering whether a shark of large size really could attack a man on board a boat, or even sink that boat. Here again, some factual stories are stranger than fiction.

On the physiological and physical level, some observations seem to me to be of interest. With regard to kinetics, we have seen that only certain species, capable of top speeds, are able to leap out of the water. This therefore excludes the slow species or the overweight individuals, and even then it is difficult to imagine a cold-blooded fish, such as the shark, making spectacular leaps like those of the warm-blooded mammals, such as the dolphins or the killer whales. It is physically and physiologically possible for some sharks to shoot their mouths one or two metres above the surface, but I do not think that they can go much higher and no evidence exists in this direction. However, it is clear that a shark heading, even at a slow speed, for the hull of a boat can shatter it like a walnut. This is often seen with killer whales or sperm whales, which are much heavier, but just imagine the kinetic energy that a 1000 kilo shark can contain, even going at only 10 knots. If in addition this kinetic energy is transmitted onto a boat hull by means of the pointed snout of a Great White for example, this is equivalent to an impact of several tonnes per square centimetre. No wooden or plastic hull can withstand such a "snoutbutt".

I have stressed elsewhere the relative fragility of the shark on impact, owing to its lack of suspensor ligaments to the organs inside the abdominal cavity and on account of its cartilaginous, rather than bony, structure. It is for these reasons that a shark taken out of the water is often doomed, even if it is returned to its natural element, particularly if it is heavy. This is no doubt also where we must look to explain the effectiveness of blows aimed at sharks to make them flee – although this is ineffective against the big specimens, which nothing can stop.

The two species which most often attack boats are the Great White and the Mako. The first is the heaviest of the sharks (up to 3000 kilos) and the second is the fastest (around 50 km/h). Both are armed with a pointed snout, as are the majority of the Lamniformes (see the directory).

Regarding the motivations for attacks on boats, Tricas and MacCosher put forward an attractive hypothesis in 1984, following experiments on the aptitude of Great White Sharks for choosing between a dead and living prey. Living animals give out a weak electric field, but one that is still sufficient to be detected by the ampullae of Lorenzini (see Chapter 3). Therefore it is quite conceivable that sharks could be attracted by boats on account of the electrical equipment found on board, in the hull or on the outside (sonar, sounder, log, etc.).

In October 1960, a fishing vessel was working in False Bay near Cape Town, and the crew were bringing in the lines, on which many fish were caught. A big shark then appeared, and began to circle the boat. The skipper ordered a halt to the fishing and went below to start the engine. However, one of his men did not follow orders and continued to fish and very soon caught another fish which he immediately pulled in towards the boat. Just as he was preparing to hoist it in over the side with a gaff, the shark swooped on the still struggling fish but missed it, and came crashing with incredible force into the boat's guardrail before falling back again and disappearing.

Although all the lines were then brought in and the boat prepared to cast off, the shark reappeared to continue its menacing circling. Suddenly it charged at great speed, fitting its jaws flush into the hull and making a hole of 45 centimetres, luckily above the waterline. A large fragment of tooth (18 mm) was later recovered from the breach, and identified by experts as belonging to the upper jaw of a large Great White Shark (known as Blue Pointer in the RSA).

Again in South Africa, in 1946 at Table Bay, a small boat was sunk by a shark, which was on the point of consuming its occupants when the latter were rescued from the brink of death by another boat. Three other vessels were attacked in the same place in the same year, perhaps by the same shark which had acquired a conditioned reflex?

False Bay at the south of the Cape is without doubt one of the places in the world where most attacks on boats are recorded. In addition to the 1960 attack already mentioned, two took place in 1942 involving a 6.5 metre shark. In 1948, another boat was almost sunk. In 1958, a Great White Shark bit the propeller of a boat which was trawling in Plattenberg Bay. In 1960, another was sunk in Saldanha Bay by a Great White, although it had to go away hungry since its shipwreck victims were rescued just in time. In 1970, six boats were attacked by Great Whites or by their "cousin", the Mako.

In 1974, in one year alone, Danie Schoeman saw his boat attacked three times in False Bay. Different sharks were involved each time, since Schoeman always succeeded in hooking or killing his aggressor. In all, he was attacked five times, probably a world record, to the point where we might wonder whether his boat exhibited particular characteristics such as the electrical equipment or electric fields that excite sharks' curiosity.

In most cases, however, the boats attacked were actually fishing at the time, and one cannot be too careful in these circumstances when big sharks are seen circling in the surrounding area.

Again in the RSA, in 1977, four fishermen went to sea in a big motor boat equipped with two outboard engines – only to suddenly find that there were five of them, when a Great White Shark landed in the boat with an arabesque as spectacular as it was unexpected. Even before the shark could be killed, one of the men was seriously injured and had to be lifted off by helicopter with a crushed pelvis and a burst bladder. The motor boat was then towed back to port, where the dying shark was finished off after having destroyed nearly everything on board.

At Christmas 1946, Harry Lone moored his boat at a Gladstone jetty and went to lunch between two fishing trips. When he returned, he was surprised to discover a 180 kilo shark noisily occupying the cockpit. The animal had obviously leapt into the boat for some unknown reason, and Lone had to call on the services of a friend to kill the unwanted passenger with his fishing harpoon; although the shark still had time to bite, twist, break and crush everything it could, making poor Lone's boat unserviceable, though the latter nevertheless got a lovely catch without any great effort.

If Makos are specialists in this type of unprovoked aggression, they are even more so when being fished on a line. It is not rare for them, seemingly exhausted at the end of a long struggle, still to find sufficient energy to leap suddenly into the boat.

Tiger Sharks and Whaler Sharks (Dusky Sharks) are also well known for attacking boats, especially when they are roughly treated by their occupants. In December 1935, the Norton brothers were cruising 150 metres from the Victoria bank (in Australia) aboard a sturdy boat 5 metres long, when they caught sight of a large 4 metre shark swimming alongside them about a metre beneath the surface. They immediately thought of the shark which had been troubling the fishermen on the pier for several weeks and which had devoured a seal the week before. A. Norton saw before him a unique chance to settle the problem and lost no time in fixing the haft of a dagger to the end of an oar. Leaning out over the water, he hurled his improvised harpoon with all his strength towards the shark's back. The water immediately became stained with the shark's blood, but the shark reacted in a way that Norton had not foreseen. Instead of fleeing, it came closer, passed under the boat, and suddenly rose up, almost lifting the boat completely out of the water and nearly capsizing it. The manoeuvre incidentally also enabled it to rid itself of the harpoon that was lacerating its flesh. By no means overawed, Norton let fly with a second shot, which again hit the target but with no better results than the first one. This time the enraged animal took a run-up, and removed a piece of the ship's stem measuring 50 centimetres by 25; water immediately engulfed the boat, and it started to sink. Within a few minutes the two men found themselves in the water, with nothing to float on and nothing to defend themselves with, and they waited, resigned, for the fatal punishment which could not be long coming. Oddly, the monster did not come back, and help arrived. In fact the shark never appeared again, probably mortally wounded by the dagger blows or by the violence of its own attack, unless its congeners, attracted by the blood escaping from its wounds, had finished it off in their own way.

The instinct of aggressive self-defence is inscribed in the genome of all animals to varying degrees, even in the most primitive fishes, and the preventive measures against sharks certainly appear double-edged when this is taken into consideration. I remember firing a 44 magnum at a blunt-nosed shark of about 3 metres which was circling at the surface around our boat on the Pacific coast of Mexico. I did it solely because we had a metal hull, and because our boat could not be overturned by such a shark. I hit it at least once on the head, but it showed no aggressive reaction and disappeared soon after. It is impossible to quantify the percentage of aggressive reactions in proportion to reactions of fleeing, but the best thing is to think hard before wounding a shark. It should also be noted that the idea of "graded counterattack" is foreign to the shark, which instead adopts the law of "all or nothing" and in general is incapable of adapting its riposte or its attack to the size of its target.

So it was that, on 5th April 1953, again in Australia, a shark transformed itself into a veritable torpedo against a metal boat, without any provocation that could have justified such violence. The boat had only just got under way when the shark crashed into it with such power that three of the boat's ribs were broken, and shipwreck was only just avoided. The shark appeared groggy after the impact, and disappeared under the water without anybody knowing whether it had survived. Whether it was attracted by the metal hull, the fish flesh being used as bait, or by those fish still alive on board, remains unknown.

Curiously, it seems that attacks on boats take place in particular in places where very few attacks on swimmers (surfers, divers, etc.) are reported, and, vice versa, as if the sharks had a quota of attacks to be made on the human species and a quota on his means of transport on the water!

In April 1946, Nardelli was fishing with his son in Spencer Gulf in South Australia, a place well known for attacks on boats, when he saw an enormous shark 6 metres long seize the line. Instead of fleeing or "sounding" (diving deeply) as usual, the shark seemed immediately to identify the source of the line and rushed directly at the boat. It tore off the rudder and sent it flying into the air with a flick of its tail, then kept attacking it, tearing it apart like a mad dog.

In the same gulf in 1871, Leslie Harris was fishing with his son Tony from a small smack, when a Great White Shark suddenly spurted from the surface and crashed down heavily on the gunnel of the little boat, which started to take in water. Tony tried to push the shark back with his bare hands, while the shark attempted to slip inside the boat by thrashing its tail, all the while chattering its teeth in sinister fashion. Tony finally managed to repel the big beast, which fell back into the water with a dull thud and immediately slid under the keel, which it set about removing. Tony began to row towards the coast and safety, but it was unfortunately too late for his father, who died of a heart attack even before reaching the shore. This is the only known case in history of a human death directly attributable to the aggressive behaviour of a shark, without the victim having suffered any injury or bruise.

Sharks have always taken an interest in certain parts of boats. It has long been known that the oars or paddles have a particular attraction for them, and more than one rower has found himself left with the handle in his hand when the rest of the oar has been broken, pulled off or cut clean through. Others, taken by surprise, have been jostled in the boat, or even somersaulted overboard by the savage abruptness of the attack. Presumably the noise and the luminosity of the water churned up by the oar trigger the sharks' curiosity, as do the leather strengthening pieces that are often seen on wooden oars. In the past, when propellers and rudders were reinforced with sheets of leather, many attacks were focused on those underwater parts of boats. I think that here, too, an electromagnetic phenomenon of low amplitude, but sufficient to stimulate the sharks' Lorenzini organ, may be involved.

Coppleson recounted, some twenty years ago, how the old schooner *Rachel Cohen* was put in dry dock in an Australian port. The ship's wooden hull was protected by hundreds of leather patches which were completely covered in algae and barnacles, but also perforated by the mouth of an enormous shark, whose lower and upper teeth had remained embedded.

Even a boat without any injured fish or bait on board can attract the dangerous curiosity of sharks: for example, canoes or kayaks, perhaps because of their metal parts, their colour and the sun's reflections. In 1954, all the competitors in a canoe race at the mouth of the Hawkesbury river in New South Wales were very closely pursued right up to the finishing line by sharks of all sizes. In 1936, Slaughter was training for the national rowing championships in a scull on the Brisbane river. Suddenly he heard a strange lapping noise behind him. Turning around, he saw nothing, but a short time afterwards the boat abruptly bumped against something and almost capsized. He headed for the bank, and only then caught sight of the shark in close pursuit, just behind the rudder. The scull sank before getting to the bank, but Slaughter managed to reach dry land without being attacked. When he recovered the scull, he found a dozen shark teeth embedded on each side of the boat and the keel was punctured for over a metre.

On 1st February 1955, on the Parramatta river in Sydney, a rowing eight was busily training when a shark leapt from the water right by the middle of the boat. The number 2 felt his oar slip out of his hands, and the shark, thrown off balance in its trajectory, fell back again ten centimetres from the skiff instead of dropping right in the middle of it. The shark's tail smacked against the side, and a wave half filled the boat. The men got away suffering only slight stress and a broken oar. A few days later, 500 metres away, a scull was jostled by a shark. Barry Court, the only occupant, had time to estimate its length at 3.5 metres, but did not hang about and hastened towards the nearby bank, the shark still hot on his heels. For a further twenty minutes, the shark circled around as if lamenting its lost prey.

In February 1929, a shark of a more playful disposition decided to take charge of a canoe with three oarsman on board. He pushed the team from Glenelme pier (South Australia) to 100 metres offshore. The men were awaiting with anguish what was to follow when a motor boat came to their rescue.

As we can see, attacks on boats claim relatively few victims, but they do not always end only in stress. Among the detective mysteries concerning sharks and boats, there is one particularly macabre one going back to 1951, in Queensland, Australia. A superb yacht was discovered floating on the Fitzroy river, with not a single living soul aboard. However, the *dead* body of Dr Joske, a doctor from Adelaide, *was* found, alone on board the boat. He had been disembowelled, and one of his legs was missing. His body was lying on the deck in a pool of blood. There were no experts around at the time, but nevertheless it was concluded that his terrible injuries could only be the work of a shark which must have jumped aboard and left again.

DETERMINED ATTACKS

Among the savage attacks in which the shark demonstrates a deadly determination, the case of Zita Steadman is striking. On 4th January 1942, Zita was joyfully participating in a picnic at Egg Rock, just at the entrance to the port of Sydney. She had arrived there with her friends in a motor boat and after picnicing they decided to go for a swim to cool off before going home. It was three o'clock in the afternoon, the hottest time of the day. All but two of them dived in, and Zita swam a little farther out than the others, although she still remained in shallow water. Her friend Burns called to her not to go out too far, and she was just beginning to come back, when she suddenly disappeared under the water. All her friends saw was an enormous shark furiously beating the water where Zita had been only a few seconds before.

Burns grabbed an oar from the boat and fought his way over to the shark in 1.5 metres of water. He began to hit it furiously, but it continued to secure its prize with such ferocity that its body emerged from the water in a bubbling of foam. The oar that Burns hurled at it did not have the least effect and it gradually dragged Zita into deeper water.

Burns had no idea of Zita's injuries, but he was spattered with blood and in desperation he rushed into a rowing boat, hoping to use its hull to push away the big beast. However, this attempt was no more successful. Finally he caught sight of Zita's long brown hair floating at the surface and managed to grab her. He prepared to pull with all his might but, to his great surprise, Zita's body rose easily to the surface. He could now see the monster clearly below him, and realised that it had not let go but had literally cut the poor woman in two and kept one part for itself.

Eleven months later, on 26th December 1942, a few hundred metres from the spot where Zita Steadman had been so savagely killed, 15 year-old Denise Burch was bathing near the bank in about 1.5 metres of water. Suddenly she was seized by her legs by an animal that would not release its catch until it tore the limbs from its victim's body. Denise Burch died even before being brought back to shore.

The savage determination of these two attacks was certainly the work of the same shark, for it is rare for a shark to maintain a single hold. Even when it has decided to attack in order to feed, in other words to tear away flesh, it does so in successive attacks, rarely in single ones. Another feature common to both these attacks concerns the time of year when they occurred: January and December. All the attacks perpetrated in the port of Sydney (eighteen) have taken place between 25th December and 2nd February, including ten during the month of January. The influence of temperature on the frequency of attack is known, and the months of December and January are indeed the hottest in the southern hemisphere, but no doubt an additional factor (perhaps linked with temperature) exists which is associated with sharks' reproduction periods. During these periods, even the most docile animal species become fierce as soon as their offspring is threatened. These months therefore might correspond with the arrival of the young in shallow waters. However, as the habits of sharks are not well known this is unconfirmed, and it could just as well be a case of the first months following phases of winter rest, or even of a change in feeding habits in accordance with the migrations towards the coast of mullet, salmon and other fish.

Whatever the reason for these periods of attack may be, there is no doubt that it is closely linked to the temperature of the water, which governs metabolism, hatching, migrations, etc.

The ferocity of an attack can also find expression in a repetition of successive charges. The undersea hunter Terry Manuel was the victim of one such attack in January 1974, in Australia, right before the eyes of his companion John Talbert who was waiting for him in their boat. The shark repeated its charge four times in succession, until finally keeping its hold around Manuel's waist. Horrified, John Talbert could do nothing but retrieve the top part of his friend's body, the rest was never found again. These macabre examples prove that a shark can cut a person in two at any level, and that all those injuries that do *not* involve transfixing are so only through the "will" of the shark and in no case through the mechanical toughness of a human body.

It is sometimes said that sharks are "timid", but when they have decided to attack in order to eat, nothing can stop them. Certain nautical or aquatic events have been the theatre of inexorable dramas played out before the eyes of a crowd of onlookers. On 15th February 1930, a national dinghy race was taking place at Middle Brighton in Australia. Once the race had ended, at six o'clock, Norman Clarke, aged 19, dived into deep water some four hundred metres from the beach, from a pier on which many spectators were still standing. He was the only one in the water. When he was only three metres from the pier, he suddenly raised his arms, screamed, and disappeared.

He then reappeared at the surface with one leg in the mouth of a shark. He was seated across the shark's snout, pounding it on the head with both hands. He disappeared once more and was never seen again. The scene had unfolded right before the horrified eyes of a hundred or so people on the pier. All agreed that the size of the shark was about 5 metres. One witness had even caught sight of the shark before it attacked, but his shouts had not been heard by Clarke, and in any case the latter would probably not have had time to get back to the pier steps. The feverish activity surrounding the race had no doubt attracted the shark, which had waited for a favourable moment. Clarke could only have provoked its curiosity by diving noisily from several metres up. Many examples exist of victims being attacked just after they have dived from a boat.

On 4th March 1956, at Port Phillip, Australia, John Wishart and five other surfers rested 250 metres from the beach after a surfing competition. The shark was already close to the beach, and passed two swimmers as it headed straight for Wishart. He was suddenly sucked under the water, and the startled onlookers saw him reappear frantically hitting the shark. Wishart vanished again beneath the surface, and his body was never found. These two attacks were probably the act of a Great White Shark.

On 19th August 1967, Robert Bartle was doing some underwater fishing in Julien Bay in Western Australia, 300 kilometres from Perth. Bartle was the national champion in this sport, which is highly rated in the region, a region where attacks are rare on account of the cold water that flows up from the Antarctic Current from the west. However, Bartle's fate was seriously to shake the illusion of safety supported by this cold current. Robert was about 750 metres offshore when a big shark seized him at a depth of four or five metres. His friend Warner witnessed the attack, and shot a dart into the shark's head in the area which he estimated to be that of the brain. It was too late, poor Bartle was cut in two, and then the shark turned its attention on Warner, circling closer and closer around him. Warner beat it off with his .20 centimetre gun, which unfortunately was not loaded. The shark disappeared with the bolt firmly embedded in its skull, drawing the 115 kilo line attached to it as it went. Due to the savagery of the attack, it seemed likely that a Great White was responsible, all the more so because it is the only one not to keep within the limits of the temperature belt. But Warner spoke of a white membrane, or of what seemed to be a white membrane, that moved horizontally and apparently covered the assailant's eye. This last point does not favour the Great White, which does not possess a nictitating membrane, even though it may expose part of its white eyeball when it rolls this back at the point of attack. All efforts made to capture the killer were in vain.

REMARKABLE ATTACKS

Any shark attack is in itself dramatic and regrettable, but certain additional circumstances, conditions or parameters can make it even more tragic, or on the contrary comfort those optimists who think that every serious accident could have been even worse.

On 27th October 1937, at Coolangatta in New South Wales, Australia, it was half-past five in the afternoon and several men were bathing 200 metres off the beach. Among them were Norman Girvan, Jack Brinkley and Gordon Doniger, who were swimming in an area immediately above an underwater channel hollowed out in the sand by currents. The three friends were fooling around and joking about sharks and then decided to get back to the shore. It was at this moment that Girvan shouted to Doniger: "Quick Don, a shark's caught me." Doniger thought that he was still joking, but when Girvan lifted his arm there was blood everywhere. "It wouldn't let me go. It had my leg," he said. Doniger swam over to Girvan, realising then that his friend was terribly shaken in every sense. But just as he reached him, Girvan was torn from his arms and the enormous shark surfaced right next to him. Girvan moaned: "I'm gone. Goodbye," and almost immediately the killer dragged him underwater.

Doniger called Brinkley to come and help him, but, just as Brinkley was starting to swim towards them, a shark attacked him. Joseph Doniger, who had seen the first spectacle from the shore, launched himself into the water towards his brother. As he was swimming, he saw Brinkley being attacked by a second shark, slightly smaller than the first, and succeeded in catching hold of him beneath the chin and started to bring him back to the shore, but the shark charged again and Joseph felt terrific jolts shaking his friend's body. He now had a perfect view of the shark, which according to him was at least 2.5 metres long.

Norman Girvan had disappeared, but pieces of his body were later thrown up on the shore in the days following the attack. Brinkley was taken to hospital in Coolangatta with his left arm torn off and the whole of his left flank gashed. He was given a blood transfusion and then operated on, but he died that evening. The next day, a female Tiger Shark measuring 3.6 metres and weighing 385 kilos was captured not far from the spot. When its stomach was opened up, undigested human arms and legs were found in it, and it was possible to identify Girvan's right hand from a scar. Dr Birch, who had examined Brinkley's injuries, stated that they appeared to have been made with a razor, without the usual characteristic tearing. The conclusion was then drawn that the principal target had been Girvan and that there had probably only been one shark involved in this tragedy. It was only when returning to attack Girvan that the Tiger would have brushed against Brinkley with its sharp-edged fins, thus explaining the linear form of his wounds. It is in fact quite exceptional for two sharks to attack at the same time, apart from during feeding frenzies, which call for very particular circumstances, and apart from cases where there are several sources of blood in the water (after shipwrecks).

It is very probable, then, that Joseph Doniger was mistaken in his judgement regarding the two sharks. I have explained elsewhere how the leading edges of sharks' fins are sharp and cutting, accounting for their ability to cause fatal injuries to a man stock-still in the water when the shark charges a great speed towards its prey (see photos in sealed section).

Another comparable attack took place, this time in fresh water, 20 kilometres upstream from the mouth of the Maria river (Australia). This was in November 1947. Three brothers were diving in the murky waters of the river just in front of their house, when Rupert, aged 13, suddenly started to yell. The water boiled up furiously around him and he swam rapidly for the bank, leaving a reddish wake behind him. Almost immediately afterwards his brother Edwin, 12 years old, screamed and disappeared beneath the surface. As soon as he reappeared, his older brother Stanley grabbed him and tried to drag him towards the bank, but in vain. Suddenly Edwin appeared to have freed himself, but in fact the shark had just released him by cutting off his leg at knee level. Edwin died on the beach in his brother's arms. Meanwhile Rupert had managed to reach *terra firma*, with a deep gash from the thigh to below the kneecap, but he recovered from his injury. It is certain that Edwin was the object of the attack, and that his brother was wounded during the shark's charge by one of its fins.

On 24th December 1934, in the waters off Brisbane, 2 kilometres from the sea, three brothers and sisters were preparing to dive from the pontoon between their house and the landing stage. Joyce dived first and her brother, Roy, applauded her while decreeing that he could do better. But just as he dived Joyce began to scream as she saw a fin coming straight at her. She beat the water with her legs and then felt a pain in the region of her knee, like a cut. Then her brother disappeared in a swirl of foam, and she herself just had time to reach the pontoon ladder a metre away from her. Their mother, who was watching the scene from a distance, began to run towards them, while their sister, Kathleen, made a dash for a nearby boat and reached Roy, who had just reappeared screaming and splashing about with the energy that comes of desperation. She just had time to touch him before he was again dragged beneath the surface. And although they searched for hours, he was never to be found again. In this last case, again the target was a single child, and the survivor was only unintentionally wounded by the shark.

On 29th December 1961, Margaret Hobbs, aged 18, and Martin Streffens, aged 24, stood motionless in the water, 5 metres from the shore in a depth of 1 metre, not far from the town of Mackay, Australia. They had been flirting about in this way for about twenty minutes when the young girl was abruptly torn from her friend's arms, in a shower of water that prevented him from seeing whatever it was. Martin succeeded in grasping Margaret's body, but the killer was stronger. When the two poor wretches were rescued, it was already too late for Margaret, whose right arm was cut at shoulder level, the left forearm above the wrist, and the right thigh ravaged down to the femur. Martin had to have his right hand amputated, as it had been lacerated by the shark as it kept after its victim. Here again, the "accessory" victim was injured only because he found himself in the path to the chosen prey. Potential rescuers should be aware that the risk of going to help a victim who is already wounded is very limited, for in the great majority of cases the shark, if it attacks again, will harass the same victim. The number of rescuers who become victims of their own courage is infinitesimal, which is easily explained when we remember the multiple and extremely sophisticated neurosensorial means which the shark has at its disposal for finding its prey again at short range, even in the dark.

Iona Asaï was an Aboriginal island pearl-fisher to whom his companions attributed the power of communicating with the gods. The unique adventure that he experienced in 1937 did not diminish him in the eyes of his companions, indeed the opposite. He himself later wrote: "During 1937, one Friday just before 11 o'clock, I dived for the third time and walked along the bottom towards a small mound. The shark was on the other side; initially I couldn't see it and it couldn't see me. I saw a stone like a pearl oyster on the north side and, when I turned around, I saw a shark two metres from me. It opened its mouth. I had no chance at all of escaping from it. It came and bit me on the head. As my skull was too hard, it then swallowed my head and placed its teeth around my neck. And then it bit me. When I felt its teeth sink into my flesh, I put my hands around its head and crushed its eyes until it let me go, and then made for the boat as best I could. The skipper hauled me into the boat and I passed out."

Three weeks and 200 stitches later, Iona developed a small abscess in the neck region, from which emerged a Tiger Shark tooth that must have been at least 3 centimetres long. Nineteen years previously, Iona had been attacked off Cairns, subsequently confirming the biblical congruity of his name: Iona is the local name for Jonah or Jonas.

In 1913, another fisherman by the name of Treacle also almost lost his head in the Torres Strait, when he dived from a boat directly into the open mouth of a Tiger Shark. However, this shark eventually released its prey, intact. Treacle, frozen with fear, made no attempt to defend himself and did not move, and had the miraculous good fortune to be released. These examples confirm that the shark does not choose one area of the body above another, but only that part of its victim that is the most convenient to seize. Where man is concerned, this means in the vast majority of cases, the legs, the knees, the hips, and, exceptionally, the head.

The two men chose very different means of defence. Asaï chose the incisive method and was lucky to be able to reach the eyes, which are one of the shark's vulnerable areas. Treacle, for his part, preferred to play dead. Such examples are too exceptional to derive from them any method of defence or to say whether immobility is preferable to aggression. Let us say that the most extraordinary chance can sometimes get us out of the worst situations, and we must always remember that.

Surfers are often victims of the curiosity of sharks, even when their position on the board does not make them look like any pinniped. On 10th December 1977, young Kim Pearce was sitting on his board, 400 metres from the shore near the mouth of the Qolera river in the Transkei (South Africa), waiting for a favourable wave. The water was clear and its temperature barely any more than 17°C. It was 10 o'clock when Kim was almost thrown from his board by a very violent bump. He looked around to see who or what could have produced such an impact when he caught sight of a shark bearing straight down on him, "at a terrifying speed". This time the impact was on the side of the board, and Kim only just managed to keep hold of it. The shark then began to swim in circles, coming ever closer, and Kim, hypnotised, did not think to lift up his legs which were hanging in the water. The shark suddenly approached on the left and seized both legs at thigh height. The impact was so violent that Kim was projected almost completely out of the water. To his great surprise he was released by the shark and felt no pain. He managed to hoist himself back on to his plank and headed for the beach. A lot of blood was flowing into the water, but the shark had vanished, as if Kim had not been to its taste. His two legs were seriously lacerated around the knees, and the left leg became gangrenous and had to be amputated above the knee. The bite on his right thigh measured 25 centimetres across and could only have been inflicted by a shark with a large mouth, capable of seizing the two legs at the same time. The surgeons were unable to provide any helpful information as to the depth or the shape of the wounds, so it was an examination of the surfboard that was to identify the attacker. The depth of the tooth marks recalled the dentition of *Galeocerdo cuvieri*, the Tiger Shark. Some Tiger Shark teeth were placed in the imprints and, as those of a 3 metre specimen fitted perfectly, it was concluded that a Tiger of around 3 metres was very probably involved.

This type of violent, but single and not relentless attack is actually more frequent than the repeated attack resulting in certain death. It is what we may call an "investigative charge", in the course of which the shark, on contacting the object that intrigues it, will bring into play most of its sensory organs, biting being the most incisive of these means of investigation.

In the Tuamotu islands, some real-life stories surpass the finest of legends. One of the most extraordinary was experienced at the beginning of the century by a diver who remains anonymous. This diver was known as an aggressive and not always very sociable man, and for the most part it was only underwater that he remained placid until the day when, as he was calmly picking up pearl oysters, he ran into a formidable creature. The shark made a wide circle around him, and then went into the attack. The diver took refuge in a coral cave, but the shark turned away and then reappeared right behind him. Seeing no other means of flight, the diver managed to cling desperately on to the shark's back and thrust his fingers as far as he could into the animal's gills.

Panicking at this unexpected turn of events, the shark rushed to the surface at great speed, emerged from the water, performed a brief arabesque, dived back towards the bottom, then came back again to the surface with the diver still gripping its gills. Raving mad, the animal made a rush for a slightly submerged reef where it found itself half out of water. The diver tumbled off his sinister mount and regained dry land without assistance, but with his body covered in lacerations and blood from the shark's rough skin and from colliding with the sharp coral. He reflected on the matter for a few minutes, became mad with rage, returned to his stranded torturer, and gave it a phenomenal punch on the snout. The shark's mouth immediately opened at prodigius speed and tore off his hand.

In July 1926, a young sailor, 26 year old Tony Madison, went overboard from his boat. As soon as his disappearance was notified, the captain made a half-turn, and by a miracle Madison was spotted at 2 o'clock in the morning. Two buoys with lights were thrown to him, and the crew saw the castaway fighting desperately in the water. When he was brought back on board, his two legs were mauled by shark bites, and his face was dotted and torn from the bill blows of large birds that had ceaselessly harassed him during his two-hour ordeal.

On 16th May 1949, at Broome, Australia, Mary Passaris, aged 22, was attacked by a shark which tore off her left forearm as she was bathing. Five days later, when she was recuperating in hospital, a 2.75 metre shark was captured near the site of the attack. It was one of the carcharinids (also called "whalers", on account of this family's liking for whale flesh). It weighed 180 kilos. On opening up its stomach, Mary Passaris's arm was found, intact and not yet digested. On one of the fingers was a ring, which Mary now wears on her other hand. The notable points about this attack are the shark's probable attraction to the metallic glitter of the ring, the state of non-digestion and non-putrefaction of the limb even after five days (already discussed in the section dealing with shark's digestion), and in addition recovery of the piece of jewellery.

About 20 years ago, off Antibes, in France, a swimmer was severely injured in the shoulder by what was very probably a shark. News reports of this attack were discreet, and nobody attempted any serious investigation. Another suspected similar attack took place at the same time in Corsica. This is not to talk seriously about such a risk existing on the Mediterranean coasts of France, but simply to underline that nothing would be impossible in that area owing to the species present in the Mediterranean, the ease with which sharks migrate and similar attacks proved at the same time in the Adriatic. It is known that certain boats are followed over very long distances by sharks, waiting at appointed times for refuse bins, kitchen scraps, meat that has gone off, etc. to be emptied over the side. If these solitary wanderers do not encounter any differences in temperature, it is quite conceivable that they could come up as far as the Mediterranean, to Marseilles, Corsica, Rome and so on.

Some attacks carried out by sharks may not bring them any luck, when they swallow things indiscriminately. Thus one of the shark's natural enemies, apart from the giant squid of the deep and the big *Porosus* crocodile, is the little porcupine-fish, a fish bristling with spines which inflates itself when alarmed. These porcupine-fishes when inflated and stuck inside a shark's mouth can easily cause the latter's death by suffocation. "Shanghai Bill" was also a victim of his own hunger, and in a way that was even more eccentric. At the time of the great sailing ships, many ports harboured in their waters sinister mascots which were baptised with exotic nicknames. The most famous were "Port Royal Jack", who guarded the entrance to the port of Kingston in Jamaica, and Shanghai Bill, who haunted the waters of the port of Bridgetown, Barbados. Shanghai Bill had already gulped down numerous sailors when he met an end unworthy of him. One day when a big sheepdog fell into the waters of the port, Bill rushed at the intruder and succeeded in taking just one mouthful out of it. However, the mouthful was enormous, and full of very long, very dense hairs. The hairs became entangled in Shanghai Bill's teeth and in the end he died. This is doubtless the only time that a dog has killed a shark. In normal circumstances dogs are favourite prey of sharks, and a succession of attacks on humans by solitary marauders have often been accompanied, or preceded, by attacks on dogs which were swimming near the shore.

COURAGEOUS VICTIMS

On 12th March 1960, at Aldinga south of Adelaide, Brian Rodger was attacked by a 3.5 metre Great White Shark while doing some underwater fishing. The shark tore away his left leg in its first attack, but, as the leg remained attached by a shred of flesh, returned immediately to finish the job. Brave Rodger had time to aim at it with his gun, fortunately loaded, and let fly with a bolt that hit it just behind the left eye. The shark did not seem to acknowledge the blow but broke off its attack and went away into the depths. Rodger was alone more than a kilometre from the coast and only a miracle could save him. He decided to use the tension cord of his harpoon gun, now useless, to try to make a tourniquet to put on his thigh. He then realised that his left hand and his forearm were also badly injured. He nevertheless managed with his right hand to position the tourniquet to stop the blood flow and so reduced the risk of attracting all the sharks in the surrounding area. Fortunately he was familiar with first-aid techniques and the vascular pressure points.

He then swam towards the shore with the energy of a man who knows he has every chance of dying, at the same time trying to control the bleeding from his forearm by compression in the region of the fold of the elbow. He was expecting the final charge at any moment, but continued to battle against a fate that he refused to accept. Thus he was able to reach a small boat occupied by two other underwater fishermen. One of them immediately jumped overboard to help him haul himself into the boat, and a three-hour operation finally saved him. There was in fact no miracle, only a fierce determination to survive, a good athletic condition, and some knowledge of physiology. He was certainly lucky that a second attack did not follow the first, but a Great White Shark generally moves away before returning, subject to the other sharks in the vicinity allowing him the time to. It is probable that of the two sources of blood the sharks chose that of their own kind first – the wounded Great White – for we have seen that they are provided with organs for detecting the vibrations of distressed fish. The vibrations emitted by Rodger were no doubt not as alluring as those of the shark he had harpooned.

On 10th December 1962, 23 kilometres away, another shark of the same size attacked young Corner while he was underwater-fishing. His friend Phillips quickly joined him and hit the shark with all his might, but to no effect. When the animal at last released its victim, he left him only with his right leg. Phillips managed to get Corner on to his board and they embarked on a nightmarish return towards the shore. Throughout the journey the monster followed them slowly at a few metres, watching each of their movements, all set to charge once more at the board which was dripping with blood. When they reached land, Corner was dead. If Phillips had reacted quickly and put on a tourniquet, it is probable that his bravery would not have been in vain and that his friend would not have died. Too many people think that terrible injuries can be arrested only by doctors, whereas actions of absolute simplicity can be practised by anybody and are enough to save the victim's life.

On 10th December 1963, again at Aldinga, and again during an underwater-fishing trip, but during a supervised competition, Rodney Fox was ferociously attacked by a Great White Shark that seized him in its jaws at chest level and propelled him in this way for several metres. Fox felt considerable pressure and could not breathe, but no particular pain. He realised that he was in the monster's mouth and tried in vain to strike it with his free arm. The shark shook him several times in all directions and then slightly released its pressure. Before it could get a better grip, Fox managed to free himself and rolled on to the shark's back. The latter then saw the fish that Fox had harpooned, and immediately swallowed it before diving. It dragged behind it the wretched Fox, whose waist was attached to the fish by a cord. It was not until a dozen metres down that the cord finally broke, and Fox resurfaced in a puddle of blood. An observation boat was on the spot immediately, and Fox was saved after a three-hour operation, resuming his favourite activity five months later. His sang-froid allowed him to analyse the event as it took place, and to take advantage of the most propitious moment to escape from the stranglehold of the terrible jaws. If one examines the photos of Rodney Fox's wounds (see sealed section), it is clear that the shark did not intend to remove flesh and contented itself with an "exploratory" bite. A shark with jaws of this diameter measures more than 3 metres in length, and can, if it chooses, exert between its jaws a power of several tonnes per square centimetre, which the rib cage of a human being would not be able to stand up to.

A victim's courage can show itself after an attack as well as during it, when he returns to face the circumstances in which he was attacked a first time. Henry Bource is one of these courageous victims. In November 1964, Bource was photographing seals in the waters of Percy Islands off Queensland when a Great White rushed at him out of the blue and removed his left leg above the knee. Immediately assisted by his comrades, he was hoisted on to the boat and a tourniquet put on him. A few months later he resumed his career as an undersea photographer, fitting a flipper to his stump. Some few years later this flipper was torn off by another shark, but Bource continued to think that "sharks do what nature has programmed them to do."(see sealed section).

HEROIC RESCUERS

It was five-thirty in the afternoon on 7th May 1959 when Albert Kogler, an 18-year-old student, decided to go bathing with Shirley O'Neill, a girlfriend of the same age. They had just spent a few minutes in the sun on Baker Beach, not far from the famous Golden

Gate bridge, and ran into the sea. They did not know that they were going to be the participants in one of the best-known attacks in the United States. Not only the one farthest north on the west coast, but also one of those in which a direct witness showed the greatest courage.

Shirley O'Neill recounts: "We had been in the water for about fifteen minutes and we'd gone out around forty metres when Albert said to me: 'Don't go any farther, it could be dangerous.' We remained standing for a moment in the water, talking. We were about to return to the shore and I was looking towards the Golden Gate (eastwards) when I heard him scream. I turned towards to him and then saw this great big thing hitting the surface of the water. I didn't know whether it was a tail or a fin, but I knew that it had to be some sort of fish. There was a real seething in the water, and I knew that he was fighting with this fish which must have been very big. It was then that he yelled out: 'It's a shark... get out of here.'

"I started swimming very fast for the shore. I did a few breaststrokes, and then suddenly thought: 'I can't leave him here.' I was scared and I didn't know what to do, but I knew I couldn't abandon him like this.

"I did a half-turn and swam a few strokes towards him. He was still screaming and I thought that the fish must be eating him alive. The howling was dreadful. All I could see was blood in the water. He yelled: "Help me, help me." (The young girl succeeded in catching hold of the boy's hand to pull him towards the shore.) But, as Shirley O'Neill continues, " his arm was almost detached from his body, and I had to put one of mine around his back to try to reach the shore."

The spectacle had unfolded before the eyes of sergeant Leo Day, from the top of the observation tower perched on the cliffs overhanging the sea. "I have never seen such an example of courage; the water was frothing with blood around the boy, who was screaming and waving his arms, beckoning someone to go back. It was then I saw the girl swimming rapidly towards her friend defiant of the danger, taking no notice of the warnings."

Without any help, with superb determination and an almost superhuman effort, she struggled for twenty minutes and brought him almost to the shore before a fisherman threw her a line and pulled them up to the sand.

Shirley O'Neill did not leave her companion until the ambulance arrived, praying that he would live, but unhappily he died a few hours later. President Kennedy awarded her the young American people's medal of valour for her act of heroism.

The high latitude where this attack occurred, outside the belt where temperatures are favourable, is explained by the relative geographical protection of San Francisco Bay against the cold current passing offshore.

A much more recent case took place on 13th February 1988 at Mtunzini in Zululand in the RSA, where an onlooker as much heroic as experienced saved the life of his comrade who was destined for a certain death.

That day, 15-year-old Belinda Van Schalkwyk and her friends were surfing near the mouth of the Mlalazi river, in a place that was not protected by anti-shark nets. The estuary is the only one on the north coast of Natal that is open to the sea throughout the year, thereby allowing intense breeding activity by fish, owing to a higher temperature and the estuary's relative protection against predators. On that day intense activity had been noticed within the fish shoals, but this had not worried the young surfers. After two hours surfing, Belinda decided to lie on her board for a rest. Being frightened of sharks, she had taken care to stretch out on her board so that nothing protruded over the sides. This was not very difficult on a board of 193 centimetres, with big black and white stripes. The only thing hanging over the side was the green and yellow strap connecting her ankle to the board. Lying flat out on her belly, head resting on her arms, she was relaxing without moving when she felt a fierce jolt on the back of the board. Thinking that one of her companions had bumped into her, she turned around to check that there was nobody behind her. Just at that moment a second even more violent blow threw her off her board. The shark immediately bit her on the left thigh, three times in succession, as if trying to get a better grip. She screamed, splashed about and hit the shark as it shook its head violently, trying to lure Belinda under the water. All the other surfers rushed for the shore, except André and Philippe Thévenau. André helped Belinda to climb on to her board while Philippe started to swim towards the beach, pushing her. One year earlier André had taken a first-aid course which had included what to do in cases of shark bite. "Her femoral artery was pulled away and the blood was spurting out in fits and starts into the water. When I went to feel the wound to find a pressure point, I realised that there was no longer any thigh and no knee," André recalled. "In these circumstances I couldn't arrest the flow of blood until I finally managed to place my finger directly in the artery, which stopped the bleeding."

André and Philippe thus swam towards the beach, their surfboards on either side of Belinda's, André keeping his right hand in the wound and his finger in the vessel. A big wave threw all three of them up on the beach. André laid Belinda head-down on the sand, and used his two fists to compress the vessels, while Philippe raised her leg to reduce the bleeding. Belinda began to feel faint but they shouted at her to stay awake. They tried to hide her terrible injury from her, but even so Belinda saw what remained of her leg. "I'm glad that I saw it at that moment," she was later to say, "for I knew straightaway that nothing could be done." Fifteen minutes later, an employee from the neighbouring park arrived with a shark-attack pack issued by the Natal Shark Board on all the beaches of Natal. A doctor who arrived shortly afterwards inserted a drip, and they waited thirty minutes for Belinda's condition to stabilise. It was only after this time that Belinda was carried to an ambulance, without at any time having lost consciousness: "We had to wait a long time for the ambulance, and as it had no oxygen and no plasma we had to wait for the other one. I started to feel pain for the first time, especially when I was moved and when I had to wait at the hospital before the operation."

After a transfusion of six units of blood, her leg was amputated below the hip and she left hospital a few weeks later.

The attacker could not be identified from the evidence given by Belinda, who had seen only a grey shape under the water, but some tooth fragments were found in her wound and formally identified by Robert Wilson of the NSB as belonging to a Zambezi Shark, *Carcharhinus leucas*. The shape of the wounds, moreover, favoured this type of shark, and not a Great White Shark (see photos in sealed section).

This attack is a model one for more than one reason, which I shall summarise in chronological order:

- Attraction of sharks to objects at the surface, especially if they have contrasting colours (in the present case the attacker could not see Belinda and she was neither injured nor menstruating).

- Savagery and determination of some attacks; this attack was manifestly dictated by hunger, since Belinda felt three successive grasps, as if to seize a larger portion of flesh, and since the bite was followed by tearing and removal of flesh. The shark really "ate" part of its victim.

- Importance of active defence once the attack has begun; there is no doubt that, if Belinda had not hit her assailant, the latter would have continued to make an easy meal out of a defenceless prey.

- Usefulness of simple actions to stop bleeding and of the "shark-attack pack". One of the most important actions is to bring blood supplies to the injured person and not the reverse, for this latter would considerably increase stress by hindering the period of haemodynamic stabilisation (haemostasis). Reassuring the casualty is also a deciding factor in reducing stress, which can bring about fatal shock. The NSB recommends thirty minutes of stabilisation before transporting the injured person, a period which seems reasonable, if, of course, one has a drip at one's disposal to compensate for the bleeding.

- Usefulness of identifying the aggressor; subsequent to this attack, a net-fishing operation was organised which enabled several Zambezi Sharks to be captured, including some of a size probably corresponding to that of the one that had attacked Belinda.

As in many attacks, we find in the environment the accumulation of one or more risk factors:

- the water temperature that day was 29.6°C;

- the mouth of a nearby river was a permanent risk factor;

- an undersea trench caused by local currents runs along the Mtunzini beach, about 1.7 metres deep, between the beach and the surf, and this has been seen to be a favoured retreat of sharks.

The other parameters were not significant: the sea was calm and without debris, the wind light and the visibility in the water about 2 metres.

This extraordinary story gives me the chance to clarify a point concerning vascular anatomy which runs contrary to many preconceived ideas, and may one day help any potential first-aider confronted with such circumstances. The neurovascular bundle which passes through the whole length of the thigh is made up of the femoral artery and the femoral vein. The artery is the largest vessel of the system apart from the abdominal vessels, and its delivery is enormous. If the femoral vessels are ruptured, the artery goes into spasm by reflex in virtually all cases, owing to the muscle fibres that line its walls. Were it otherwise, the 4 to 5 litres of blood contained in the body's system would be emptied out in only a few seconds and death would be instantaneous. It is in fact the femoral vein that remains wide open for it does not contain any muscle fibres, and it is this vein that must be sealed. Venal bleeding is less cataclysmic than arterial bleeding, but it requires swift intervention, like that which saved Belinda's life.

A PAINLESS TORTURE

Whatever the mode of attack and however serious the injuries suffered by victims, there is one factor that is as remarkable as it is extraordinarily constant: the total absence of pain. This fact is regularly reported by victims, without any scientist ever, to my knowledge, having tried to provide an explanation. An explanation is necessary, however, when we see the gravity and the heterogeneity of wounds: clean or jagged lacerations, deep perforations, detached limbs, flesh torn off, multiple cuts, bone fractures, etc.

Having often had the opportunity to help the recovery of polytraumatised individuals with open fractures and wounds, I have noted their very considerable pain, and there is rarely total analgesia for comparable injuries.

First we must eliminate the possibility of the presence of an anaesthetic in the dentition of sharks. As everything in the animal world has an adaptive objective aimed at the survival of the species, I find it hard to see where the advantage could be for sharks not to cause their victims any suffering. Whether they suffer or not, they are in any case incapable of escaping from this formidable predator once it has bitten them. Animals as vulnerable as leeches benefit by injecting a local anaesthetic under the skin of their host so that the latter is not aware of anything, but the shark? If this anaesthetic existed, it would not only be an extraordinarily effective one but also impossibly fast-acting. Sharks' teeth certainly come into the study and classification of wounds, not because they emit an extremely powerful local anaesthetic; but rather during the study of the secondary infection of wounds, which is sometimes fatal, due to the putrefaction of organic matter lodged between sharks' teeth.

If anaesthesia does not derive from direct contact between the shark and its victim at the point of biting, it could still be induced by proximity to the shark, perhaps through some electromagnetic method. The transmission of pain is a depolarising wave that follows the main nerve stems in a centripetal direction (towards the centre), from the wound to the nerve centres, and it is conceivable that this transmission may be disrupted by an electromagnetic emission or field. Magnetotherapy is based on the fact that a magnetic field of 1500 gauss or more applied to the skin is capable of short-circuiting the passage of a pain message. This is how the application of small magnets on the skin allows a rapid rehabilitation of sprained ankles or minor traumatic nerve injuries. The shark does not have any magnets on it, but we might suppose that its mysterious lateral line is also an emitter (many marine mammals, and no doubt fish, are capable of using echolocation). The route of this lateral line around the fish's body is in that case comparable to a single turn of a coil which releases a low-frequency induced current. We could allow an analogy with the phenomenon of ELF (Extremely Low Frequency), a top-secret working hypothesis of NASA in the 1980s, a phenomenon that could bring about a temporary stunning of the nociceptive nerve impulse. This remote interaction was used, in a quite different sphere, aboard minesweepers in the last war, which had a demagnetisation ring encircling their decks to counter floating mines. This purely personal hypothesis is perhaps incorrect, but I think that an explanation must be provided, to account for this temporary anaesthesia.

If such a line of research is not acceptable, then the explanation must come from the very nature of the injuries inflicted. The speed at which they are inflicted, within a few tenths of a second, is perhaps a cause of immediate non-sensation, but how do we explain there being no pain in the minutes that follow?

Road-accident victims often suffer loss of limbs, and this can occur extremely rapidly, in something like a few hundredths of a second, but intense pain generally sets in very quickly as long as the injured person is conscious.

Could the multiplicity of wounds in the same limb segment saturate the pain message? Many cases exist of injuries to lower limbs where the sciatic nerve has been cut. Understandably the distal wounds below this cut would not be affected from the pain point of view, but how do we account for loss of feeling at the same level as this rupture instead of syncopic pain?

Another explanation could be the salty, hypertonic, aqueous surroundings in which these sudden wounds occur. The depolarisation wave accompanying pain accompanies the transfer of ions, and in this situation these could be disturbed. It would be interesting in this connection to compare the pain felt during attacks in fresh water in relation to that from attacks in the sea.

I have long thought that the very acute stress caused by being attacked by an invisible monster in an environment that is not our own was sufficient to suppress pain through the very rapid secretion of anaesthetising endorphins. I am sure that this explanation is valid in the few tens of seconds following the attack, but much less so as regards the instantaneousness of the anaesthesia. The truth is perhaps a combination, with temporary anaesthesia caused by the shark itself, followed by a hormonal discharge that takes over.

Whatever the reason may be, it will be interesting one day to discover for certain the explanation accounting for the painless nature of the terrible injuries inflicted on man by the shark: from this a better understanding of the mechanism of pain will no doubt be deduced, and, perhaps, a revolutionary method of therapeutics.

POST-MORTEM OF AN ATTACK

After an attack has occurred, it is always very interesting to be able to identify the animal responsible. The first detailed investigation made by an experienced observer goes back to 1952, when Barry Wilson was attacked and killed by a Great White Shark at Monterey Bay, California. The study made by Rolf Bolin constitutes a prototype of all the parameters that must be taken into account, from the climatic conditions, the turbidity of the water, the chronology of the attack, the rescue efforts, the possible causes, to photographing the victim's injuries, and taking an exhaustive medical description of these injuries. This investigation marked the beginning of a scientific methodology for the analysis of shark attacks. The detail of the factors to be taken into consideration appears right throughout this book, and I shall not therefore make a list of them in this section, stressing only certain methods of identification.

When an object has been bitten before or during the attack (boat keel, surfboard, diving suit etc.), it is interesting to note its colour, its general size and the noise it was emitting, in order to identify the stimuli that may have attracted the shark. Where a sailboard or surfboard is involved, its thickness, the general diameter of the tooth marks and their orientation, shape and depth must be noted. These simple observations can suffice to identify the shark, and in any case to provide an accurate idea of its size from the diameter of its mouth (see diagrams). A tooth fragment will sometimes be found, which is generally enough to identify the order to which the shark belongs, if not its species. Sophisticated methods of radiography at low voltage and long exposure time have enabled tooth fragments measuring a millimetre to be found inside the synthetic foam from which surfboards are made.

Bone X-rays allow a diagnosis to be made as well. For example, the case of Michael Mely, whose hip X-rays showed deep notches on the iliac crest. The appearance of the lesions on the soft parts did not permit a categoric diagnosis; on the other hand, these narrow impressions 3 centimetres deep could only be the work of the Ragged-tooth Shark, whose teeth are not chisel-shaped but elongated

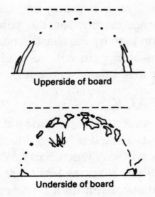

Upperside of board

Underside of board

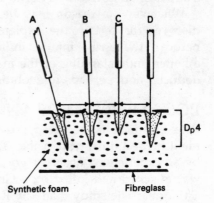

Synthetic foam   Fibreglass

The pattern of tooth marks on K. Doudt's surfboard (attacked in Oregon) enable us to confirm that a Great White was involved, and to estimate its size accurately.

X-ray of Michael Mely's iliac crest showing the shape of the bone lesions left by the shark's teeth

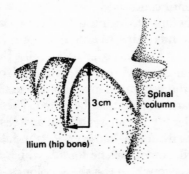

3 cm   Spinal column

Ilium (hip bone)

Placing needles into the imprints provides an accurate idea of the length of the teeth, remembering they are not parallel.

1. Simple wound from investigative bite.
2. Wound after sideways shaking of the head to tear.

156

and pointed like daggers. The abdominal wound showed circular rows of small triangular lesions without any tearing, proving that the bite had been a simple investigation dictated by curiosity and not by hunger. If hunger had been the motivation, there would have been stripping and tearing of the flesh, and the appearance of the lesions would have been different.

The savagery of the attack and the appearance of the injuries are sometimes deciding factors for prejudging the species involved. For example, on 24th December 1960, 25 year-old Petrus Sithole was bathing in the muddy waters of Margate (south of Durban). The water temperature was 24°C. About about four thirty in the afternoon, some witnesses saw Sithole swimming strongly right above a channel three metres deep, about 70 metres out from shore. They suddenly heard him scream and saw him beating the water with both hands while the whole upper half of his body shot vertically above the surface. His cries ceased as he fell back face down in the water which quickly became red with blood. Two men, who were swimming nearby, rushed for the shore without attempting to help him. However, waves brought him close to the land, and a man then ran to fetch him when he was waist-deep in the water, but the poor wretch was already dead. His left leg was cut clean through just below the hip, and the femur had been shattered by the savagery of the attack. The right leg had been cleanly severed at knee level and the lower extremity of the femur was sticking out, intact, through the muscle tissue. A skin abrasion, three centimetres wide, crossed the chest and abdomen, probably caused by the impact of the shark's snout, or by any other part of its rough-skinned body. The remarkable feature of Sithole's injuries was the clean nature of the bites, resulting in immediate amputation of both legs. Such injuries are characteristic of the Great White Shark or of carcharhinids such as the Zambezi Shark (*C. leucas*). A Ragged-tooth is also capable of amputation, but the sides of the wounds are shredded and torn (see the shape of the teeth in the directory). The autopsy found no tooth fragment; but a post-morten X-ray revealed the presence of minute pieces of teeth stuck in the lower part of the right femur. A microscopic examination of these fragments as well as comparison with whole teeth in the collection at the Durban Oceanographic Research Institute identified the assailant. One of the fragments corresponded almost exactly with the fore lateral teeth of the upper jaw, and the other fragment with the fore lateral teeth of the lower jaw of a Zambezi Shark. From the size of these same fragments, the shark's size was estimated at 3 metres and its weight at 200 kilos. It is sometimes impossible to tell the difference between

C. carcharias and *C. leucas* by simple macroscopic observation, and in South Africa an index file is now available listing what I shall call "scan signatures" particular to the teeth of each species of shark. The thickness of the different tissues in a tooth is in fact characteristic for each shark species.

Another biological investigation is based on the blood of sharks. It is common, in fact, to find blood on the objects surrounding the victim at the moment of attack. There is, of course, the wounded person's blood but, in addition, also that of the shark. The shark's teeth are subjected to a severe test at the time of an attack, and they rest on a fleshy tissue which bleeds easily. If you examine photographs in which the wide-open mouth of a shark seems about to swallow you, you will notice that many teeth have blood-tinged bases. Small specimens of this blood are frequently found inside impressions left by the shark's jaws on surfboards, on boat keels etc. Some people have hoped to be able to identify the species that have attacked from these blood samples (Lea and Miller, 1985), and the method is an appealing one bearing in mind the enormous progress made in the field of haematology, and in particular in the typing of a very large number of specific sub-groups. No doubt we shall soon be able to identify a shark from analysis of a trace of its blood.

On 20th January 1989, a 19 year-old boy was surfing at Isipingo just south of Durban. It was six o'clock in the evening when Sudash Sarjoo was grabbed by his left thigh by a shark, which immediately dragged him below the surface. Under the water Sarjoo was able to observe the pointed snout of his assailant and its "white eye". Released by the shark, he was brought back to the beach by his two companions and taken to hospital. He was suffering from two deep crescent-shaped wounds, one of which penetrated his femur (see photos). These wounds seemed to have been caused by the teeth of the upper jaw, and their shape indicated that the teeth were finely serrated. Sarjoo had thought that he recognised a Great White, but the wounds left by the upper jaw were no more than 10 millimetres deep, which is incompatible with the Great White and with the Zambezi Shark, whose lower teeth are over 20 millimetres in length (for a jaw with a diameter of 25 centimetres, as in this case). Only a Tiger Shark of around 2 metres in length can account for such injuries in a diameter of 25 centimetres, with lower teeth which are no longer than 10 millimetres. This diagnosis is, besides, perfectly compatible with the white eye that Sudash Sarjoo thought he saw, and which corresponded to the closing of the nictitating membrane characteristic of all Carcharhiniformes at the moment of attack.

SOME IMPOSSIBLE STATISTICS

The only statistic established so far for the total number of shark attacks in the world was calculated by David Baldridge over twenty years ago, and covered all attacks up to 1967. This study concluded that there have only been 2000 attacks, fatal or otherwise, over two centuries. Baldridge was only prepared to record those reports, verbal or written, which he regarded as 100% reliable, which explains his derisory and, I have no hesitation in saying, totally unreal figure. While the 2000 cases are a statistically valid sample for looking at the circumstances in which shark attacks occur and develop, in no way do they constitute an exhaustive historic list. If it were otherwise, I should for my part certainly not have written a book about so negligible a risk. It is obvious that the risk remains marginal in relation to the population as a whole – the person of sedentary habits who goes swimming ten times a year in the English Channel will never be confronted with "shark danger". By contrast, all those who have regularly swum, dived, hunted, sailed or surfed in tropical waters know how the heart misses a beat when a shark's fin appears ploughing through the surface of the water, or when its magnificent silhouette shoots out of the blue deep. And what about those who have been shipwrecked in tropical waters and who have practically always suffered from the less than friendly presence of sharks?

As I have pointed out in the section concerned, there are around 50,000 victims of shipwrecks per year throughout the world, at least half of them occurring in tropical waters. It is very difficult to estimate how many of them are taken by sharks, for the dead cannot speak, and many bodies are never found. But I think that the order of magnitude is from several tens to a few hundreds according to the year (very probably several hundreds for the *Dona Paz* alone in 1988). This is a long way from the averages established by Baldridge.

Baldridge himself reveals an incredible bias in his sources of information: 90% of the attacks that he listed took place in English-speaking countries. Would sharks be Anglophiles to such an extent that only Commonwealth subjects suit their tastes? To the extent of ignoring all Africans except the South Africans? All the Latin-Americans? All the Asians except those in Singapore? All the Mediterraneans except Egyptians? All islanders in the Pacific not living in Hawaii?

If we analyse the distribution of these attacks, we find that they are concentrated on the coasts of North America, South Africa and Australia, almost totally excluding more than 75% of tropical coasts, all of which, however, are known to be frequented by many dangerous species. For the 20,000 or so kilometres of South American coast, only two attacks are mentioned, although this continent has its coasts on either side of the equator within the entire intertropical zone.

Everybody in the 1990s knows that the Republic of South Africa is one of the most vulnerable regions in the world to shark attacks, but who has ever heard this subject mentioned for Mozambique, Pointe-Noire in the Congo, Madagascar, Angola, Ethiopia, and all the French-speaking countries, all of whose coasts are even hotter than those of the RSA and their waters rich in sharks of all species? As far as attack statistics are concerned, these countries do not exist. And yet I know that Mozambique, for example, is the theatre of very real dramas every year, in numbers even more significant than in its rich southern neighbour. The waters are the same, the coasts and the currents are the same and the sharks are the same. There are no bathing resorts protected by nets, but swimmers and divers all along the coasts, all potential victims. The presence of sharks has for ever been embedded in the country's culture, but the danger they represent is accepted as one of the risks of the occupation of fishing, and nothing more.

This fatalism goes hand in hand with a misinformed or uninformed media; with a different view of the value of human life; with quite a different idea of what a "scoop" is; with often total geographical isolation; and with a political desire not to have disappearances of this kind publicized. It needed carnage like that which happened in 1960 for the Mozambique newspapers to devote a few columns to sharks. In August that year, a boat capsized at the mouth of the Komati river, and the 49 passengers ended up in the water; 46 of them were mutilated by a pack of sharks. Nevertheless, if we refer to official statistics, no attack has ever been recorded in Mozambique.

On the other side of Africa, I have had the opportunity to fly over the whole coast of Senegal and Gambia on numerous occasions, at low altitude, and I was staggered to find each time, whatever the time of day, the presence of hundreds of sharks a few metres from the shore. The shadows of the sharks stood out very clearly in the surf, floating at the surface in groups of two or three, and measuring between one to three metres. The sandy beaches stretch unbroken for hundreds of kilometres, and the sharks are everywhere, and in even greater density in the vicinity of villages. These villages exist on fishing, and thousands of natives of all ages go into the water every day. Again according to the statistics, there are no attacks recorded from Senegal, which is of course impossible. All the game-fishing enthusiasts who go specially to Dakar (in Senegal) to devote themselves to their hobby know very well that sharks are very numerous there, and not among the most harmless.

One need only listen to those involved describing the incidents or accidents of which they are regularly victims. I have seen tens of professional fishermen sporting spectacular scars or partially amputated limbs, not only in Senegal but also in Angola, in the Congo and in Latin America. But all those survivors who express themselves in French, in Portuguese, in dialect or in Spanish are not included in the statistics. On the island of Goré, at Dakar, there is a slave museum where an entire room is given over to shark attacks, which were in the past enough to stop slaves escaping.

I lived in Angola, in 1981, in the dark days of Marxism-Leninism, and one of my rare distractions was the sailboard which I had managed to bring in my luggage. After a few weeks, I was sure that there were sharks in the waters around Luanda. I had been pursued by fins on several occasions and had easily outdistanced them, but I was not happy about their presence, particularly as the water was cloudy, and the beaches had been deserted for years. I tried to find out more about the potential risk, but it was absolutely impossible to get any information at all from the authorities. Nevertheless, the few fishermen regularly brought in dangerous sharks, and the hundred or so cargo boats, which had to wait off-shore for several months before being unloaded, inevitably attracted sharks through the refuse emptied overboard. I learned later that a number of swimmers and windsurfers had been killed by sharks around Luanda, but of this, once again, there was not a word spoken at the time.

Madagascar is another country where there are no official records of attack. Ask any Frenchman who has lived for even a few months on this huge island right out in the Indian Ocean, and he will immediately tell you of a number of cases confirming that sharks are everywhere. From the tragic story of the little girl who was carried off in fifty centimetres of water at Tamatave before the eyes of her parents, to the tale of the "Tulear oxen". The cargo boats anchoring at this port in southwest Madagascar were obliged to unload the cattle they brought from Europe one by one, using a crane, with big harnesses slung under the animals' abdomens. This brief trip was not always to the liking of the unfortunate livestock, which panicked and occasionally became unhooked. They were then inevitably devoured by the sharks, which had developed the habit of lurking under the wharf.

The island of Réunion is another example of missing statistics, whereas the history books and other written and verbal accounts report numerous accidents. About fifteen years ago, one of my friends was bathing with a companion who, like him, was doing his military service on Réunion. He heard his friend cry out some distance from him, saw him struggle and then disappear. By the time he reached the spot where he had vanished, he found nothing but a few traces of blood in the murky water.

Before the last war, a very well-known French politician was seized by an enormous shark on a Réunion beach, and his injuries were comparable to those suffered by Rodney Fox in 1963. He never used his spectacular scars for electoral persuasion, but his story shows that nobody is immune to attack, whatever their status! However, it is important to note here how many divers and how many scientists have been attacked by the subjects of their observation all over the world. It is the frequency of exposure to risk that creates the real danger.

Baldridge's statistics also completely omit the victims in the Second World War. Although it is perfectly reasonable to estimate these at several thousand if we remember that the wrecking of the *Nova Scotia* and of the *Indianapolis* alone claimed between 500 and 2000 lives attributable solely to sharks.

On a historical and geographical level, we ought to be surprised at the very limited number of attacks recorded in the Mediterranean, since in ancient times Herodotus spoke of thousands of soldiers and sailors being "seized and devoured by monsters" off Mount Athos when the Persian fleet was destroyed in a storm. In AD 77, Pliny the Elder mentions divers involved in "frantic combats with sea dogs which attack the back, the heels, and all the pale parts of the body"; this last detail, which could surely not have been the product of imagination, does indeed identify these "sea dogs" as sharks. And yet the official statistics list only 18 attacks in the Mediterranean up to 1963, and none since. All the big predators exist in the Mediterranean, apart from the Tiger Shark and the Bull Shark: Great White, Mako, hammerheads, Blue Shark, Copper Shark. It is very likely that it was the opening of the Suez Canal that was behind the increase in the shark population. Although as long ago as 1889, an English surgeon reported the case of three victims of shark bites which he had had to tend in Port Said. The northernmost attack was in Yugoslavia in 1934, and caused the death of Agnes Novak. The majority of attacks are officially in Greece and Italy, almost always in summer, but only a single case is given for the whole of Maghreb. Once again this is implausible, and the lack of exactness is no doubt attributable to the geographical isolation of the victims and/or to "media black-out". Imagine, for example, how little the Gaddafi authorities must think of a victim of these "terrorists of the sea".

I know that, in at least two Mediterranean countries, the authorities have effectively hushed up certain attacks which occurred in renowned tourist waters. When we recall the economic impact of the Italian "red mud", it is easy to understand why the officials are not keen to dramatise events over which they really have no hold, but it does not necessarily follow that the scientific authorities are as naive or as lacking in curiosity.

PREVENTATIVE MEASURES AGAINST SHARKS

A defenceless man in the water confronted by a threatening shark can survive, there are methods, even if they are last-ditch ones. The victim of shipwreck in a lifeboat in tropical waters is clearly less vulnerable, although not absolutely safe, and for him the means of avoiding provoking the curiosity of sharks come under the simple heading of common sense, subject to knowing a little of their physiology.

The increasing popularity of watersports exposes more and more people to the risk of attack. Holidaymakers from temperate or cold regions are going in ever-increasing numbers to tropical latitudes, and every year there seems to be some newly designed craft on the water. Ever more rapid or more original, sail or power-driven, with or without propeller, this vessel is added to all the others to bring more and more people into contact with sharks, and the statistical risk of attack increases accordingly. In the rich countries and particularly in those most vulnerable to attacks such as the Republic of South Africa and Australia, nets have been put in position and are very effective in protecting those areas delimited by them. Outside these privileged perimeters, careful thought and caution are the only means of limiting the risk.

In earlier chapters of this book, while examining specific case histories, I have already suggested some measures for the prevention of shark attack. In this section, I shall bring all these suggestions together and add to them the fruit of scientific research and other, pragmatic observation. I shall distinguish three basic categories of prevention:

• prevention through education, or an understanding of sharks and from this the attitudes to adopt;

- prevention through preventive equipment, individual or collective;

- close-range protection, passive or aggressive.

PREVENTION THROUGH EDUCATION

Do not entice the potential attacker

Knowing the shark's hypersensitivity to odours, and its acute attraction to certain ones that evoke its favourite prey animals, refrain from swimming near wounded fish or any living thing bearing a wound, even a superficial one.

If you injure yourself when diving in tropical waters, come out again without delay

Some swimmers have been savagely attacked after scratching a limb on a branch of coral. I have already pointed out that women should not go bathing during their menstrual period in risky waters.

If you go underwater fishing, bring each catch up.

Either deposit it in a buoy at the surface, or, better, into the boat accompanying you. Cut off the head of the fish, otherwise there is a risk that its last dying movements will be discerned by the shark through the bottom of the boat.

About fifteen years ago, I was a diving instructor in Mexico in a region with a very rich animal life. On one dive I did not notice that one of the girls from the diving "stockade" had attached to her bikini top a transparent plastic bag containing some pieces of fresh tunny to feed to an enormous grouper that we had spotted the day before. I only became aware of it when, in 30 metres of water, another grouper ripped off her top together with the plastic bag that it was after. As I could not get the animal off her and it was earnestly attacking her chest and causing her to bleed, I stabbed it with a dagger. It disappeared, leaving behind it a trail of blood, followed immediately by a Black-tip Shark that had been swimming near us. It, too, was attracted by the odour emanating from the plastic bag but also by the vibrations emitted by the attacking grouper. When this latter was wounded, other sensory stimuli made it change its prey: the distress vibrations of the grouper, the smell of blood, and the sight of the fleeing fish.

I have cited in this book several examples of underwater fishermen who died because they kept the fruits of their hunting on their person; this is a mistake tantamount to suicide in tropical waters.

Besides the smell of the blood of the injured fish, another very powerful stimulus exists for the shark, and this is the odour of a "fish under stress", even if uninjured. Albert Tester was able to demonstrate this beyond all doubt in 1960 in Hawaii. He placed intact fish of the same species in two tanks. In the second one he incited them to panic by shouting and beating the tank's walls with a stick. He siphoned the water from the first tank into a shark pool, and the sharks reacted very little. However, the water from the second tank, which had contained the panicked fish, instantly provoked an intense hunting activity, some sharks even going so far as to bite the outlet tube of the siphon. A fish pierced by a harpoon bolt or trapped by a hook therefore attracts sharks by the smell of its blood, "the odour of stress" and the vibrations its emits.

Avoid throwing organic debris into the water

Shipwrecked people in a dinghy will have to avoid throwing organic debris over the side, or do so only periodically in hermetically sealed plastic bags thrown as far away as possible.

Rubbish tips bordering the sea, fishing grounds or whale processing plants have always been sources of attraction for the sharks within a radius of 10 kilometres, and are therefore dangerous areas. A few years ago, a luxury hotel was inaugurated with great ceremony in one of the French Caribbean islands. A few months later it had to shut through lack of clients, potential guests being scared out of their wits by the sharks which cruised around in front of the hotel or, more accurately, in front of the fishery just next door.

Do not stimulate the sensory organs of sharks

When you are familiar with the extraordinarily high-performance functioning of the shark's eight sensory organs, the behaviour to adopt or to avoid so as not to stimulate them is logical. For the man in the sea, the most dangerous point comes when the swimmer or diver reaches the surface before hoisting himself aboard his boat. It so happens that it is also the most tense, when eyes above the surface watch in anticipation but cannot see what is going on below, but the reason for the danger does not lie in this. The shark is always attracted by any object whose shadow stands out against the surface, which is always very bright, even at night.

A man at the surface no longer looks like a human being. All the shark sees is a shadow of increasingly reduced bulk the closer it finds itself to the vertical. Breaking out from this shadow are two mobile, spindly appendages (the legs) splashing about against the luminous background of the surface, and the shark is naturally attracted by these shapes which remind it of certain prey (fish, pinnipeds, squid, etc.). Moreover, when the legs move about in line with the sun, this involuntary provocation increases.

In 1974, as an officer on a rapid escort vessel belong to the French Navy, I was appointed by the "governor" as the ship's second diver. We brought the ship to one afternoon in the Gulf of Guinea, some 160 kilometres from the coast, so that the crew could bathe in a magnificent sea as smooth as a millpond. The "master-at-arms" was keeping watch from a Zodiac dinghy, with a few lifebuoys on board and a submachine gun at his waist, ready to intervene should a shark put in an appearance. I was diving below the ship at a depth between fifteen and twenty metres, never tiring of admiring the superb black tapering shape that the enormous hull described on the surface. I also noticed the tens of pairs of little legs splashing about ridiculously at the surface, and the noise that came from all my mates who were shouting, puffing and diving. We were in several thousand metres of water, and I could just imagine a big Oceanic White-tip Shark emerging from the great blue deep and heading for this incongruous and unexpected mob of people. No shark made an appearance, but this was because the exercise lasted for barely more than twenty minutes. Added to all the vibrations emitted by the crew, there was a whole array of mechanical vibrations coming from the slow-running engines, the diesels at full power, the superchargers and the boilers, etc. All of this transmitted a very long distance and at great speed from the sheet metal of the ship right down to the ocean depths. It is enough to have experienced this just once in one's life to understand why sharks so often follow ships for hundreds of miles, and how they arrive so quickly at the scene of a shipwreck.

In dangerous waters, spend as little time as possible at the surface

If, however, you wish to swim in spite of everything, wear a pair of those tiny goggles which allow one to see perfectly in the water. This way you can detect the approach of a shark and keep an eye on its movements. If you are a diver, do not stay needlessly at the surface.

Dive in twos or threes, never alone, and arrange yourselves back to back if you are threatened by a shark

Scuba divers are rarely attacked, perhaps, as some people have maintained, because of the bubbles escaping from their pressure reducer, but above all I believe, because the dark colour, the bulk and the shape of a diver are more reminiscent of a dolphin or another shark than of any possible prey. A diver 1.8 metres (5 ft 11 in) tall is about the same size and bulk as a large shark, and the latter might consider that attacking such a target is not such a good idea (even if it is mistaken!).

Do not wear shiny objects

Shiny objects are by their nature attractive to sharks, and a number of attacks have been to parts of limbs bearing wedding rings, bracelets, or brightly coloured anklets. In February 1963, Rex Gallagher was trying to harpoon a big grouper as he hid in a cavern in New South Wales. As he emerged from the cavern without having injured the grouper, a shark which he did not have time to identify rushed straight at his mask. This was ripped off, the tube torn, and, under the power of the impact, shark and hunter found themselves at the surface. Wounded on the chin, on the cheek and on the nose, Gallagher got away and was never to dive again in a shiny, stainless-steel mask.

Avoid yellow or orange

I have already mentioned the stupidity of lifeboat manufacturers who persist in putting out boats with yellow rather than black hulls. Although no doubt a few square metres of black rubber stuck on the yellow bottom of a boat costing £2500 probably constitutes an unnecessary expense when one spends one's life in a quiet little backwater in northern Europe, this is light-years away from the reality of the tropics.

Noise

Certain noises can also attract sharks, as I have emphasised repeatedly, and among these are the vibrations transmitted to the surface by the rotor blades of a helicopter hovering stationary above the water. Sharks have been seen to rush on a man in the water whom they had not spotted earlier, only at the moment when a rescue helicopter came to winch him up. Underwater explosions are also a source of attraction for all the sharks which are outside the fatal zone of the shock wave. Once they are on the spot, the dead fish rewards them for their curiosity, even though the original stimulus was the explosion and nothing else. Low-frequency electronic vibrations draw them as well, explaining the bites on certain appendages on warships.

Is a man on his own more or less vulnerable than one in a group? A noisy group attracts sharks much more quickly than a silent individual who is not emitting any odour. The risk of attack for such a group is considerable, sharks appear not to be frightened by crowds, but see them as an entity full of interesting stimuli. If, on the other hand, an individual comes to be isolated from the crowd (as when surfers all go on the same wave), the alert system of the shark's sensory organs will be concentrated on him alone. The swimmer, the diver or the surfer isolated from the others is in greater danger.

All these pieces of advice, dictated by logic and especially an elementary knowledge of the physiology of sharks, will help you to acquire what Leonard Compagno calls "shark sense", that second nature which gives you an instinctive inkling of when conditions and circumstances are dangerous.

Do not provoke sharks

There are many ways of inciting the curiosity or the aggression of sharks – and for man, both motivations can cause the same result. Even the most innocent joke can sometimes take a bad turn when its subject matter is this animal with a sense of humour that is, to say the least, cutting. On 27th November 1976, Jeff Spence was bathing with a friend on Clifton beach, not far from Cape Town. Because he had drunk a few beers, was 19 years old and was happy with life, he suddenly started to imitate the young lady whose screaming introduces the famous film *Jaws*. Screaming and pretending to be struggling, he shouted to his companion: "Look out, it's going to eat you, too." Playing the same game, the latter pretended to rush away from the scene of this mock drama.

A minute later, as he was swimming normally, Jeff felt a powerful shove which propelled him rapidly upwards. Feeling no pain nor any sensation of injury, he immediately presumed it was a dolphin, as there were many in the region and he knew that they often came to cause a rumpus with the bathers.

It was only when the animal released him and he found himself covered in blood, that he realised that it was no longer time to joke. His injuries were comparable to those of Rodney Fox (see the photo), the shoulder and the whole left side of the chest were torn, perforated right to the bones. Jeff swam as fast as he could in the opposite direction, expecting a second attack at any moment. His companion observed the fin heading for Jeff and then circling in the film of blood, searching for the wounded prey. Fortunately the two men were picked up by a boat just as the shark broadened its boundary of investigation and headed towards them. Jeff's wounds took several hours to stitch at the Groote Schuur hospital. However, he was as miraculously lucky as Fox that he survived a chest bite and that his attacker had "decided", in this case too, not to shut its jaws. In both cases the bites were "investigative", with no intention of removing flesh. Jeff's shark, nevertheless, was pretty well armed, measuring around 3 metres according to witnesses, and judging from the shape of the wounds was a Great White. The experts agreed that Jeff's wilful screams and movement could only have aroused the curiosity of the shark, which happened to be in the vicinity, conjuring up for it the vibrations of a wounded fish.

If you bathe or dive in waters likely to be frequented by sharks, be quiet, and make full movements and not abrupt ones, in a word, act like a fish in the water.

Bruce Wright, who was in command of a British reconnaissance unit during the last war, made a comparative study on the effect of blood and movements on the aggressiveness of sharks. He came to the conclusion that blood on its own attracted barracudas and sharks, but that it was rapid and uncoordinated sudden movements, with or without the presence of blood, that excited them much more (he did not know at the time about the movement detectors with which sharks are equipped).

In 1976, I used to dive very regularly with a Mexican friend who was considered to be one of his country's top specialists. One day when we had been exploring the magnificent depths bordering the coast of the island of Cozumel for nearly two hours, about 1 mile offshore, we saw, 50 metres from us, an enormous shark seemingly asleep on the edge of the shelf, which at this spot breaks off into a 400 metre trench. My companion signalled to me to watch what he was going to do, and I saw him approach the shark from behind. It was not until he was beside its tail that I realised the size of the animal, which was in fact moving imperceptibly to keep itself facing into the current. Around 3.5 metres in length and strongly tapering with a pointed snout and a symmetrical tail that seemed as tall as my friend, I was wondering what species it might be when I saw my Mexican bring out his dagger. I immediately thought that he had gone mad, easily imagining the fate that the monster would have in store for us if roused by a stab from a dagger, so I drew out mine just in case. Hardly had he touched it, when the shark took off like an arrow, disappearing into the depths. When I asked my friend what he had been playing at, he replied that he had "wanted to check if the animal was sick, as normally "Makos" do not stay on the bottom like that"! The provocation was enormous, especially with a Mako of that size and 30 metres down, and I still think that we were very lucky that day.

Peremptory recommendations for prudence

The following pieces of advice are not the result of scientific debate, but the fruits of hundreds of rigorous observations, in particular in the RSA.

- *Do not go too far beyond the surf.*

- *Do not swim close to underwater channels* which often cross the sandbanks between the beach and the surf. Find out from the regular visitors about the possible location of these channels.

- *Do not swim in murky waters,* especially after floods, and avoid the vicinity of a river mouth.

- *Do not swim or dive at night or at dusk.* Like all the terrestrial predators, sharks mainly feed at night. On several occasions when diving at night with a friend, each of us scanning a sector of the darkness with his torch, along with the myriads of luminescent animals, we saw several sharks, whose mouths in these conditions had nothing reassuring about them, no more than did their black eyes, which glowed in the night.

- *Do not be rough with apparently inoffensive sharks, even small ones.* A number of incautious people have been seriously injured like this because they insisted on walking on carpet sharks or other coral-reef sharks.

- *If you feel something brush against or touch you, leave the water* to check if anything has happened to you. Sharks like to "taste" their prey by lightly touching it before biting it (the taste buds on their sides give them this remarkable capacity).

- *When on a dive, never let a shark get between you and any obstacle.* Its reaction will be that of the snake, which will only attack when it is cornered or is directly attacked itself.

- *Do not think that you are safe because dolphins are in the vicinity.* Dolphins sometimes decide to kill a shark, and they manage this by swooping on them at great speed and planting their pointed snout in the fish's gills or abdomen. Their intelligence as mammals allows them to have no fear of sharks, but do not think for all that that they are going to come and protect you. Dolphins and large predators have been seen living happily together, at least for a brief time. In February 1989, north of Port

Elizabeth (RSA), I dived near an island covered with thousands of penguins, and surrounded of course by tens of sharks of all species, including the most dangerous. During the thirty minute dive in cloudy water in which visibility was no more than five metres, I heard and then saw coming close to me three large dolphins and two Ragged-tooth Sharks, the most dangerous in South Africa. As they had thousands of web-footed birds available to them a hundred metres away, I knew that they would probably not "take a little taste" of this "fish" that was almost as big as they were, and in actual fact they disappeared. Even if it had been any different, the dolphins certainly would not have intervened.

- Sharks never immediately attack a diver who has just left the surface, so *you have time to decide calmly whether to go to the bottom and come back up when the time is right*. A shark which is interested in a diver begins by circling around him, leaving and returning, and this behaviour may go on for a very long period.

- *If you have handled any fish, wash before diving.*

- *If you see a fin, it is not necessarily a shark*. It may be the fin of a swordfish, a dolphin or a porpoise (they generally leap out of the water, quickly removing any doubt), a killer whale (the fin is very big and very black) or even a manta ray (in which case there will be two "wingtips" which will move in an abnormally synchronous fashion).

PREVENTION THROUGH FACILITIES

Surveillance

Much in vogue in Australia, where sharks are virtually everywhere and where it is not possible to set up nets on all beaches, are those who act as lookouts for sharks, a job comparable to being a lifeguard. The jobs for that matter are usually done by the same people in the same observation towers, with a big bell to hand which gives the alarm as soon as a shark appears. The Australians bathe on most beaches between flags which mark off the zones being watched, and swiftly leave the water once the alert is given, calmly returning as soon as a boat has driven off or captured the intruder. As we can see, it is a method requiring public-spiritedness and discipline, qualities which the Anglo-Saxons know how to show. If the coasts of France had the misfortune to come under threat, this would probably not be the best method.

Air patrols also exist during the summer, from helicopters, from planes or from microlights, these last proving very effective. Land-based systems for long-range intruder detecting exist, and no doubt there will soon be undersea equivalents of such detectors which could be used to warn of the appearance of sharks in the vicinity of beaches. It will not be a question of infra-red detectors for cold-blooded animals, but of movement detectors, doppler, sonar, etc.

Rigid barriers

Historically, in Australia as well as in the RSA, the first protection methods were based on solid barriers, with very limited boundaries. Apart from being very expensive, they never stood up long to storms, and maintaining them was extremely constraining, even for local councils or governments. Stationary or mobile, metallic or nylon, none ever lasted more than a few years. I went bathing in 1969 at Veracruz, on the Caribbean coast of Mexico, inside one of these barriers which was made of railway sleepers and iron bars. When, after an hour, my friends and I decided to go and "challenge the monsters" on the other side of the rails, we were surprised to find that all the bars had been torn away by the sea and only the useless vertical sleepers remained.

The first fixed barrier was erected in Durban in 1907, followed by Cogee in New South Wales in 1929 and then at Sydney and elsewhere in Australia. But there were always gaps which allowed large specimens to get through, as well as channels underneath the barriers created by the currents, and some fish even swam over the top on the highest tides. The final death knell for these barriers was sounded when a bather was attacked and killed inside one of them in Queensland.

Now that there are extremely tough modern materials on the market, it will perhaps again be possible to construct physically impassable barriers based on composite plastics, especially Kevlar, which is resistant to saltwater corrosion.

The NSB's nylon nets: a remarkably effective method

Each year 1200 adult sharks are captured, from the dangerous Zambezi Shark to the Blue Shark, the Tiger, the Mako, the Hammerhead and most of the other large predators. This is the record of the Natal Shark Board in Durban, the "Miami of South Africa" sustained for more than 20 years. In the wake of the terrible series of attacks during 1957–1958, the local administrations of the province of Natal decided to put up protection nets, following Durban's example. These nets turned out to be an essential factor in avoiding ruining the seaside tourist trade, but they threatened to bankrupt the local councils and therefore the provincial administration took on the expense of providing protection. It was the woman who took over the management of the NSB in 1966 who progressively turned it into a body of unique public and scientific interest. Beulah Davis is probably the most famous woman in the Republic of South Africa. Heading 250 people who protect 450 beaches every day, with 385 anti-shark nets, she also has to manage a fleet of 45 boats specially designed to get through the surf in all weathers, and to maintain a bank of scientific data on sharks which is unmatched anywhere else in the world.

To use nets as an effective means of capturing sharks was discovered by chance in 1937 by an Australian fisherman, who noticed that the nets that he set for tunnies and other carangid fishes very often brought in big sharks, but no other fish of large size. At the time the fisherman did not know the reason for this, but we now know that it is due to the simple fact that an immobilised shark very quickly becomes asphyxiated (see *Respiration*). If it cannot manage to free itself very swiftly, unlike other fish it suffocates and rapidly dies. The net therefore not only acts as a "lobster pot" but also as a deadly trap.

The arrangement of the nets is surprising, since neither vertically nor horizontally do they constitute a continuous barrier. Each net measures 107 metres in length by 6 to 7 metres in height, and is separated from its neighbour by an open gap of several metres. The area covered in the vertical plane is 20 metres, and the net does not touch the bottom nor does it reach the surface. This means that a shark can pass over, under or around by dodging in and out of the nets, which are set out in staggered groups of five. This is what happens in fact, since a large percentage of the sharks are recovered facing the open sea direction. The disadvantage of this penetration inside the protected zones is a minor one, as it occurs mainly at night. The sharks come to hunt at dusk near the beaches, travelling at the surface; but they return to the open sea at dawn, swimming deep down, this is how it is possible for them to be captured facing in both directions. These nets are therefore nothing other than a static fishing operation, simply reducing the number of sharks in the protected zone and thus lowering the statistical risk of encounter with man. I could liken the method to that used in epidemiology, where the spread of a disease is stopped by vaccinating only 50% of the population.

The staggered arrangement of these nets is necessary to allow the currents, the objects floating beneath the surface, and the other fish to pass through. It is only at the time of the "Sardine Run" that the nets are removed for a few days. At this time, the sardines migrate along the coasts of Natal in their billions, to such an extent that, lacking oxygen, their bodies accumulate in great unbroken heaps along the beaches. This rush of sardines is of course followed by the arrival of the sharks from all over the region, and any nets that are not yet

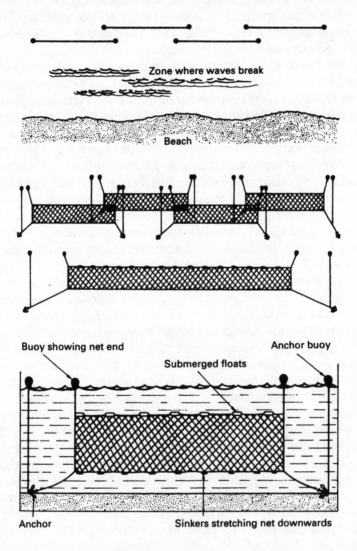

Typical configuration of NSB's anti-shark nets

taken in are then brought down by the tons of sharks caught in them (see photos).

To begin with the nets were made of green polyethylene, and then they were changed for others of black nylon. The green colour could be more easily seen by the sharks than the black, and the strength of the polyethylene was found to be halved in the area where it was knotted (the black one has a resistance of 375 kilos). This black nylon was specially manufactured for the purpose and is interwoven with blue, although this is for a reason which has nothing whatsoever to do with sharks explained to me by Beulah Davis: the local children, all conforming to the fashion of wearing black bracelets, used to go out in a boat to pull up and cut pieces of net to put around their wrists! She also told me how a few years ago a net was found with a clean hole measuring several metres across, and how the NSB inquiry had secretly come to the conclusion that it had been penetrated by a pocket submarine belonging to a power well known for that kind of infiltration.

In 1989, I was invited to South Africa by the NSB and given complete freedom to take part in and have access to anything that interested me. Of course I joined the outing to raise the nets at dawn in one of the boats which daily check, change and maintain all the nets on all the beaches. A net is brought up only when the traction on it is significant, indicating that a shark has been caught. When the shark is too heavy, divers go down to free it. When a shark is still alive, Beulah Davis insists on its being released, quite rightly considering that "it is for man to adapt to sharks, and not the reverse". I was thus able to witness the seriousness with which each captured shark is tagged and listed according to date and place of capture, before being sent to the NSB headquarters where it is stored in the cold room to await autopsy. The centre's computers thus store an enormous bank of information on the general physiology of sharks. Every week, public autopsies are performed on three sharks to inform and teach the general public.

The nets are regularly put out to dry in the open air every three weeks in order to get rid of the algae and micro-organisms that attach themselves to them. When a storm prevents their being pulled up and there is a risk that they will collapse, the beaches are put out of bounds.

The effectiveness of these nets is almost complete, since only one attack has been recorded inside the protected zones in 20 years. Outside these zones, attacks continue, for it is of course impossible to place nets along the entire 2000 kilometres of coastline.

Australia also uses similar nets, but in a less systematic fashion. The meshes are of the same dimension (30 square centimetres) but the nets are a bit longer. The results are also encouraging. Each shark captured costs the community around 30 dollars.

Barriers of bubbles

Following the theory that the low toll of attacks on scuba divers was due to the emission of bubbles that surrounded them, various bodies decided to examine whether curtains of artificial bubbles would frighten off sharks. I shall not give a chronological account of the experiments which were very seriously conducted, notably by Dr Perry Gilbert, both with air bubbles and with bubbles of carbon dioxide and other gases, since these studies proved the ineffectiveness of the method.

Electric barriers

Experiments in this field are more encouraging. Electrical fishing is still in its infancy, and perhaps the day will come when electric nets or barriers will kill or paralyse shoals of fish outright. We know, moreover, that sharks are extremely sensitive to the weakest of electric fields, and it is conceivable that strong fields could upset or frighten them. At Port Maching near Sydney, a 3.5 metre shark was literally paralysed when crossing an electric barrier; it remained motionless until the current was switched off, at which point it carried on swimming as if nothing had happened.

More elaborate experiments have been carried out on bony fish. When a current is passing between two electrodes and a fish finds itself between these electrodes, it orients itself towards the positive pole, but we cannot tell whether the current attracts fish outside the field. These methods are used commercially but have not yet been tried out in salt water, on account of the amount of current required. Better results have been obtained in fresh water, particularly in the former USSR.

Experiments made in Pretoria (by Dr Lochner) have recently confirmed that a current passing between two electrodes acted as a barrier to sharks. These experiments also confirmed, and this is important, that the charges necessary to keep sharks away did not trouble swimmers.

A cable with alternating current has been installed off the beach at Margate (Natal), forming an electromagnetic barrier. This experiment on a true-to-life scale goes back to 1988, and its results seem encouraging, even though maintenance of the cable still appears difficult. However, it could soon be a viable alternative to setting up nets.

As long ago as 1961, John Hicks perfected an electronic "sharkshocker" which seemed to work very well, and in which NASA shared an interest. Unfortunately his little dinghy was overturned by a slightly panic-stricken shark, and that was enough for the officials in the aquarium where he usually worked to ban him from continuing his experiments. He remains the originator of a small individual electronic emitter which can be fitted to a diving suit and which wards off over-inquisitive sharks. This instrument is called "Hick's repeller", and is based on an acoustic frequency specific to certain species.

Explosives

At the end of the 1950s, the South African frigate *Vrystaat* received the order to release explosive charges off the Margate-Uvongo coast, where a number of attacks had just taken place. Twenty-four explosions directly above 13 "shark trenches" produced only eight deaths, and 39 other charges in the days which followed did not kill a single shark. On the other hand, many fish perished during this exercise, and their bodies attracted numerous sharks. The authorities were happy to let the local police, who patrolled aboard a small outboard, toss grenades at all the sharks they saw at surface level. This exercise might have been amusing and was no doubt more effective than the first, however the results never seemed very clear. I have, however, read the report by Commander Fane concerning the exploding of half-kilo bars of TNT: "A large number of fish were lying on the bottom or had risen to the surface. In twenty seconds, sharks arrived from every direction, devouring the biggest fish head-first. If a piece of fish was hanging from the jaws of one shark, it was seized by the other sharks. Observation from an underwater porthole showed that the sharks fled at the moment of detonation, only to return ten to twenty seconds later. Repeated explosions did not keep them away. The sharks were capable of navigating at great speed in the dense foaming fog caused by the explosions, even through the irregular labyrinth of the coral bed." Other experiments tend to prove that these fish are much more resistant than humans to the shock wave of an undersea explosion. These results reinforce my certainty that shipwrecks are a very considerable focus of attraction for sharks, and that the explosions accompanying the majority of major wrecks are nothing but additional stimuli.

In July 1956, four men in a dinghy noticed the appearance of a shark just as a Royal Navy team was about to embark on a diving exercise. They decided to frighten it off by casting a line to which they attached two 400 gram charges of explosive. The line coiled around the shark, which had the unfortunate whim to pass just beneath the dinghy at the moment the two charges exploded. Two dead and two seriously injured men were thrown into the air and landed back on the surface among the remains of the dinghy and the shark. The problem with big explosive charges under the water is the danger they present to those letting them off and the diffusion of the shock wave. The only viable charges are those contained in cartridges which propel a projectile directly into the shark.

I have long wondered why sharks seemed so insensitive to the shock waves from explosions while other fish were so susceptible to them. The explanation is simply due to the fact that sharks do not have a swimbladder, whereas the bony fish have a bladder that tears under the impact of the explosion wave. Man is likewise very susceptible to undersea explosions on account of the diversity of his abdominal organs (they are both dense and gaseous).

On 25th April 1945, the US destroyer *Frederik Davis* was torpedoed, and the survivors found themselves lost in mid Atlantic. Several minutes after the destroyer had sunk, two depth-charges exploded on board deep underwater. These explosions were sufficiently subdued by the density of the water not to cause fatal injuries to the shipwrecked men in the sea, but they indirectly brought about the deaths of many of them by attracting, in the minutes following, an impressive number of sharks from the open ocean.

Line fishing

Following certain attacks, intensive hunting campaigns have been launched, in which every means was used: among these, fishing with spectacular lines. The attackers were often caught, but can we deduce from this that line fishing is a good preventive method? I think not, even though certain commercial fishing lines reach up to 90 kilometres in length and have 2000 hooks. The day when it is deemed that proper prevention comes only through extermination of the species, then such lines will be used.

PASSIVE ON-THE-SPOT PROTECTION

Repellents

During the Second World War, the US Army became aware that many pilots who landed in the sea or men who were shipwrecked perished at the hands of sharks. Numerous poisons were tried out, including arsenic and hydrocyanic acid. The trouble with poison is that the shark's low metabolic rate enables it to bite well before being poisoned and thus the man still risked being the first victim. The solution, if it existed, therefore lay elsewhere. Supersonic waves were tried, without effect, as well as dyes and inks. It was then that a biologist became aware of a fact Florida fishermen had long been familiar with: when a shark was rotting in their nets, not one of its congeners would show itself within a radius of several kilometres. Experiments proved fruitful: a piece of decaying shark flesh was enough to keep away all the sharks in the area, even though they habitually cannibalise the first of their own kind to be mortally wounded.

The miracle substance contained in the decaying flesh was ammonium acetate, which in water produces acetic acid. Experiments were successfully carried out off Florida using an analagous substance, copper acetate. The biologists tossed a basket of small fish into the water which the sharks attacked voraciously; a second basket followed, containing more fish and, in addition, copper acetate; the sharks beat a retreat almost immediately, circling around the bait at a respectful distance but never touching it. These tests were corroborated by others off the coasts of Australia, and henceforth allied sailors and airmen were equipped with sachets of repellent known as "shark chaser", with an acetate and nigrosine base. This material at least had the advantage of reassuring future wreck victims, but the initial enthusiasm for it was increasingly tempered as the gadget was put to use and tested in a serious fashion. Not only was the Great White singularly unimpressed by the shark chaser, but some species were even seen to swallow pieces of meat filled with repellent, which then came out through their gills like an aeroplane engine losing oil. Two hundred different products were tested, but none was effective against all species, nor ever constant in its effect (certain species are more aggressive in one region than in another). Up to 1980, no repellent hd been discovered that was really worthy of its name.

For a decade and more, research has been undertaken on the strange immunity of the sole to sharks. It is the only fish that has never been recovered from the stomach of a shark and the only one that has never been seen to be bitten. It has now been possible to extract the substance that gives the sole this immunity: and it turned out to involve pardaxine, which works on the gill area of sharks. When the sole is threatened, it ejects this substance towards the shark's nostrils or mouth. Numerous experiments have been carried out with this substance, in Israel as well as in the USA, and all have come up against the problems of concentrating, conserving and synthetically producing it. Dr Zlotleni noted that the natural product had many of the properties of a surfactant, the ingredient which in detergents breaks down surface tension and produces the bubbles or froth. A number of surfactants were tested and compared with pardaxine and it was sodium dodecyl sulphate (SDS) which proved the most effective, surpassing even the repellent qualities of the natural product. This discovery is a recent one, dating from August 1987, and very promising. Experiments done on more than a hundred sharks in the wild have all proved positive. SDS has always been present in many household detergents and shampoos, and it is a great pity that its unexpected quality with regard to sharks was not suspected sooner. For Dr Gruber, who obtained research supplies from the US Navy, this method of repellent is more effective than the electricity method, for it seems to drive away sharks for good (with the Hick's repeller, they tend to come back after being driven off).

Two hundred cubic centimetres of solution to 5 or 10% of SDS is enough to repel sharks. The product is not, however, powerful enough to be used as a "screen" with which the shipwrecked person or diver can surround himself. It must be directed straight at the shark's face. Therefore cartridges of this product should always be carried and fired at the attacker's snout by means of a remote injection system. One might also consider putting cartridges of the product inside diving suits, which would prevent being bitten twice, and also in the synthetic foam part of surfboards. So if one day you are shipwrecked in tropical waters, do not forget to take all the shampoos and detergents with you on board your lifeboat!

It is obvious then that the chemical or biochemical solution to shark attack is to be found in these natural or synthetic repellents which are non-toxic to man, and certainly not in poisons such as strychnine, which were tried out twenty years ago. The way they spread in the water was terribly dangerous for the handler, and, even when a shark did swallow a bait packed with strychnine or cyanide, it took at least thirty minutes to die.

Clothing

Ron and Valerie Taylor have experimented with suits which are like coats of chain-mail, preventing the sharpest of teeth from penetrating the skin, and it is obvious that this constitutes a good protection against sharks of a certain size. On the other hand, this protection can only be illusory against a shark of 3 metres or more, which is quite capable of crushing one of your limbs in its 3 tonnes to the square centimetre vice.

Bearing the same cautions in mind, we might also look at bullet-proof jackets which are now made up of 22 intersecting layers of Kevlar and are fully effective against a 357 bullet, without being particularly cumbersome. I think that diving suits integrating the same material would transform sometimes fatal wounds into simple bruises, and yet would not obstruct undersea manoeuvrability.

Cages

Metal cages are the most reliable method of protection, whether they are positioned on the sea bed, suspended by crane from a ship, or even mobile like those devised by Cousteau's *Calypso* team. Some are extremely flimsy, such as those which encircle every one of the competitors taking part in a traditional swimming competition in Australia. These are towed at the surface by a boat which adapts its speed to that of the swimmer, and have, it appears, warded off several attacks.

Cages are the preferred observation sites for cameramen or undersea photographers, especially when they scatter blood-stained baits to attract big sharks. It should be understood, however, that

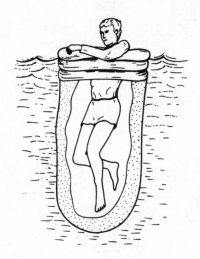

The Johnson Shark Bag.
This apparatus combines simplicity with protection against sharks and against the cold. It involves a black plastic bag filled with sea water, kept afloat by three inflatable buoys, in which the shipwrecked person immerses himself. Odours remain trapped inside the bag, as does lost body heat. The black colour and the absence of apparent movement do not attract the shark's attention. The advantage of this bag is its small size when stored before use; the drawback is its flimsiness. Considering its undeniable effectiveness and the minimal space it takes up, it should be distributed automatically to people likely to be exposed to a "shark hazard".

there is no cage that can withstand the impact of a 2 tonne shark, with jaws that have a crushing power of 5 to 6 tonnes per square centimetre. In 1990, a French undersea-photography enthusiast experienced this when, suspended in a cage off the Cape, he saw a Great White Shark in the waters. He got himself pulled up and out just before the shark, which weighed around 2 tonnes, rushed at the cage and smashed right through it. There is nothing surprising about this if we consider both the kinetic energy of a mass of 2000 kilos moving at between 5 and 10 knots, and the Great White's extremely pointed snout shape, which can penetrate the cage and brush aside the small bars like matchsticks.

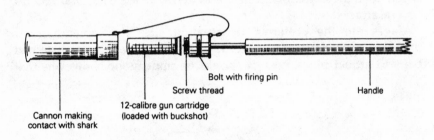

Bolt with firing pin

Screw thread

Handle

12-calibre gun cartridge
(loaded with buckshot)

Cannon making
contact with shark

Diagram of the "bangstick" or "lupara"

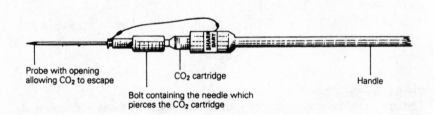

Probe with opening
allowing CO_2 to escape

CO_2 cartridge

Handle

Bolt containing the needle which
pierces the CO_2 cartridge

Diagram of the "CO_2 dart"

AGGRESSIVE ON-THE-SPOT PROTECTION

The "bang stick" or "Lupara"

Over the last fifteen and more years, offensive means of countering sharks have been developed, such as the mini-spear with exploding head which has at its tip a chamber for a 12- calibre cartridge. If the tip of the cylinder is pressed between the shark's eyes (when hauled aboard a boat, for instance), the cartridge cap bangs against the "hammer" and the projectile goes directly into the shark's brain, instantly killing it. The advantage is that there is no loss of power in the water since the explosion can occur only on contact with the fish. This system is now fully perfected with a rapid-loading exploding head for marine use (Lupara), which overcomes firing problems by effective waterproofing of the cartridge. It is without any doubt the best individual means of defence ever devised.

The CO_2 dart and the electric dart

For divers, two weapons founded on a good knowledge of the physiology of the shark are of interest in their effectiveness. The first is based on the fact that the shark's capacity to swim depends on the close relationship between the lifting power of its fins (proportional to their size), its weight and the amount of oil contained in its liver. Each species thus has its own characteristic ratios between these three factors. Should one of them be upset, the shark will no longer be able to control its depth.

This knowledge of physiology, coupled with the use of a small cylinder of carbon dioxide, is what is behind one of the most effective anti-shark weapons. The CO_2 cartridge is fitted to the end of a dart and extended with a big needle. When the shark comes too close, the diver pushes the needle firmly into any soft part of it, if possible the abdomen. It will then meet a backwards thrust, which pushes another finer needle into the gas cartridge; and the latter abruptly empties directly into the animal's body. This added buoyancy propels the shark straight to the surface and often kills it by causing a gaseous embolism. The advantage of this instrument is its simplicity, and its lack of any mechanism liable to jam in sea water. The method is an appealing one, but less easy to put into operation than it might appear. Floating anchors have also been perfected which can be planted anywhere in the shark in order to obstruct its movements. The results are usually mixed, especially with the larger individuals.

The second instrument is a 10 centimetre needle attached to a battery which is tiny (3 cubic centimetres) but which produces 300 volts. Placed on the end of a 3 metre spear, this arrow breaks off once planted in the animal's body, forming a closed circuit of arrow-shark-ocean. Depending on the shark's size, this can mean rapid death, or at least paralysis (10 minutes) in any case, time for the diver to get away.

"Striking out": a last-resort method

When totally helpless and facing a charging shark, it is obvious that your only chance is to hit it, preferably in a sensitive area, and, if possible, with a dagger. Many accounts mention aggressive sharks being driven off by their victims simply kicking or punching them; clearly, this can be effective only on sharks of moderate size, which are not too determined to attack you.

The "shovestick" is an instrument which can easily be made from a handle 1 to 2 metres long, with an adequate diameter (a spade handle rather than a broom handle) and with two or three points at its end to stop it slipping on the shark's skin, yet not so sharp that it becomes embedded. The ideal thing is to weight it so that it has zero buoyancy and is thus more manageable. This simple instrument is functional and easy to use.

I have used such a "shovestick" against reef sharks of about 2 metres, and its effectiveness surprised me. Sharks generally look for easy food, and a few good raps on the snout can make them go away. The danger caused by their unpredictability is tempered by their absence of pride. A few well-placed stabs in the eyes or the gills are often enough to drive away the attacker, often pursued by its congeners attracted by the blood.

Shouting: a disparaged method

The pearl-fishers of Ceylon are very rarely attacked and have a great belief in their method of prevention: screaming in the water. Dr Haas confirms that sharks are frightened by shouts under the water and insists that he has saved himself from several attacks by doing this. The effectiveness of these screams could come from the vibrations received by the lateral line, and/or the bubbles escaping at the same time.

The great underwater fisherman Wally Gibbins succeeded in killing a 350 kilo Tiger Shark with an explosive dart, and is used to regular confrontations with sharks. When asked if he is frightened by these great predators, he replies: "I've found the best thing to do when you are threatened: you wait until the shark is 1.5 metres from you, and then you yell as loud as you can while beating your arms and legs about. The closer the shark is, the more effective the shouting."

Sensitive areas of a shark in regions of which one should strike to drive away aggressive individuals: A. When using a blunt instrument, a knife or one's hand, the latter as a last resort only (listed in order of sensitivity): 1. The eye – this is very sensitive, the animal can be blinded if the knife is well aimed. 2. The snout is sensitive due to the concentration of various sensorial systems, such as the canals of the lateral line, the Lorenzini organs and the pit organs. 3. The sensitivity of the gills is linked with the presence of the respiratory system, with a concentration of blood vessels, muscles and branchial nerves.

B. When using an underwater gun or a weapon with loaded head: 4. Straight downwards between the eyes and through the chondrocranium to reach the brain. A double-spring gun is enough for the dart to pierce the cartilaginous skull of a 2 metre shark and, if well aimed, to immobilise or kill the animal. 5. Diagonally downwards, in the direction of the spinal cord. The best spot is at the front: avoid firing straight through the thicker skin in the middle of the dorsal median line. A well-placed dart can instantly immobilise a 2 metre shark, but the medulla is much harder to reach than the brain. 3. Gills: hit here, a shark will not be immobilised instantly, even if fatally injured; it is better for the dart to enter from the front rather than sideways on, as this will impede swimming. It is taking a risk to shoot a shark. It is a bit like "catching a tiger by its tail." If the animal is not immobilised, it will normally round on the dart and, in so doing, there is a strong chance that the diver will suffer. From *Sharks of Polynesia* by R. H. Johnson, Pacific Editions.

I have read a good number of criticisms relating to this method, but, within the framework of last-ditch remedies and remedies "which cannot do any harm", I think that the method is worth remembering. Once again, nothing will stop a monster several metres long from devouring you if it has set its heart on it.

Far-fetched methods of protection

A number of "Professor Calculus" types have taken an interest in the problem of anti-shark protection, particularly in the countries concerned and not always for philanthropic purposes. One of them installed underwater speakers in shark aquaria, through which he played various pieces of music. From this he quite seriously concluded that sharks were very fond of waltzes, but that the music of the Beatles made them flee frantically.

Another "expert", having noticed that Great White Sharks rarely swim at more than 3 knots whereas a good human swimmer is capable of 4 knots, discovered an inspired method for avoiding a shark attack, namely swimming faster than the shark.

A few years ago, the security council of British Columbia asked its ten thousand or so divers to wear pink, yellow and red clothing so as not to be taken for seals or young whales, under the impression that these animals were the same colour as normal diving suits. Another source, Dr Starle, is known to have thought that diving suits with black and white stripes, like sea-snakes, could spare people from attack. Experiments were carried out, with narrower and broader stripes, but the sharks' behaviour never changed.

CONCLUSION

The reader will no doubt have ascertained, through all the accounts contained in this book, that there is no stereotypical attack behaviour for sharks. It is nevertheless possible to derive, in a desire for an overall view, several characteristic types of approach, investigation and attack. Although total indifference from a shark is rare, an attack with severe biting is even more so. Between these two extremes, there are a number of types of shark behaviour which occur in the majority of human-shark confrontations.

A shark alerted by the presence of a live being in its close surroundings will in most cases come to make a visual inspection. If the human being does not particularly interest it and does not send it any additional stimulus, it goes its own way. If, however, the human is splashing about, glinting in the sun, and making a noise, the shark keeps watching. The best way for the shark to combine observation of a fixed point with the need to be in constant translational motion

ATTACKS: INSTRUCTIVE DRAMAS

Indifference (rare)

Approach with swift visual inspection from a distance without follow up

Approach with surveillance circling – without follow up or follow-up, contact and attack

Approach with brush-past, without follow-up (wounding possible)

Charge with collision (upwards trajectory generally)

Charge with single or double investigative bite without tearing

Charge with biting and removal of flesh (death in 45% of cases))

Multiple feeding-frenzy charge (death in 100% of cases

Different methods of investigation and attack of man by a shark in order to be able to breathe, is to circle around that point. However, this can be temporary and the shark will move off if nothing interesting is happening. If it is interested and the object does not seem dangerous, it will come and make contact to find out more about it.

This extra information can be gleaned by simply brushing against the object (contact is made with the "taste buds" and the electric detectors), and we have seen that in these cases, serious injuries can be inflicted by the leading edges of the fins and abrasions by the shark's rough skin.

This observation phase may proceed no further and the shark may disappear. However, it could also be followed by an attack, or an attack could take place immediately. The attack takes several possible forms. First the "bumping tactic", in which the shark, generally emerging from the deep, violently knocks into its victim, sometimes causing him to be pushed out of the water. A bite may be made during this intentional "collision".

The attack may also take the form of an immediate bite, which generally takes the victim by surprise. This bite is often a single one, occasionally double, and does not always involve the removal of flesh. It is what we can call an investigative bite, just as if the shark were "tasting" the object. Perhaps it is also an intimidatory bite which could be linked with a concept of territoriality, but this supposition would need confirmation. I would like to compare such an attack, which is usually not fatal, with what happens with bites from snakes considered to be deadly: in about 50% of cases, these snakes do not inject their venom, and the victims do not succumb (see my book *Survivre*, 1988).

The bite can turn out to be to the shark's liking and it will then tear its prey. Or perhaps it has decided, even before making contact, to bite in order to feed, and then flesh will be removed. Injuries from these attacks are very serious, and death occurs in about one case in every two. If the victim does not manage to escape in time and the shark's attack is dictated by hunger, the bites will be repeated and always directed at the original victim who is now bleeding (even if he is in the middle of a crowd) and generally at the same limb. Death is then inevitable.

If the attack is made during a "feeding frenzy" victims haven't got a chance, since all the sharks in the area, big and small, rush at every object in reach, dead or alive, bleeding or not, and bite them with horrifying relentlessness. This is the situation in which sharks best merit their reputation as the "killers of the seas". But in actual fact, as I have just illustrated, the shark is above of all a "man-biter" or a "man-taster", rather than a "man-killer".

6

FACT AND SPECULATION

CAPTIVITY AND ADAPTABILITY

THE DESIRE TO LEARN about shark biology has led to an interest in how they behave in unusual situations, in particular in captivity. Thus we have learned that sharks have an aptitude for learning that is both interesting and surprising for a fish. Psychophysical experiments made to examine the role of the sensory organs call for conditioned, Pavlovian, reactions in order to determine the thresholds of sensitivity to or discrimination of certain stimuli. The success of these studies bears witness to the rapid learning capacity of the elasmobranchs. Habituation, which is a simple form of learning, has shown itself to be rapid in the shark in its natural setting. It has thus been possible for numerous tests on its sensorial functions to be carried out (Myrberg, 1978; Nelson and Johnson, 1972), and as a result these animals can no longer be considered as purely primitive creatures with no capacity for learning through experience. This is of interest in explaining certain differences in behaviour observed, for example, between young and adults. The young are almost invariably more aggressive than the adults and their level of activity is at the same time higher, more erratic and more unpredictable. The reasons for such differences are unknown, but are reminiscent of what we see in the young of other species which alter their behaviour only with some experience, in other words with a certain maturity.

This adaptive capacity could explain certain serial attacks, attributable to a single marauding shark, which discovers human prey by chance and adapts to it to the extent of making it its

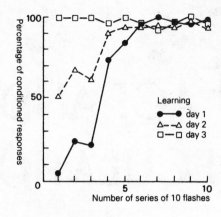

Learning curve for a conditioned closing movement of the eyelid (nictitating membrane) in a lemon shark. The training consisted of an intense luminous stimulation associated with a low-voltage electric discharge, at the rate of 10 flashes, 10 times a day. The graphs correspond to the three days of training. Note that the sharks reached 100% in conditioned responses from the sixth series of 10 flashes on the first day. This experiment demonstrates the great adaptive ability of the shark. (From Gruber and Myrberg, 1977.)

preferred, if not sole, prey. It is obviously not a question of deliberate determination but rather one of the diversion of the feeding instinct.

In 1963, when Clark was taking stock of the shark species which had been successfully kept in captivity, he arrived at some rather discouraging results. Although more than fifty species had been kept for more than seven months in captivity, many nevertheless died the following year, and it had never been possible to keep any truly pelagic species or any known to be dangerous. The species encountered most in captivity were the Sand Shark, the Cat Shark, the Leopard Shark and the Nurse Shark of the West Atlantic. The general opinion was, rightly, that not only are sharks difficult to capture and to transport but, once in captivity, they often refuse to feed themselves and very soon die (Gruber and Koyes, 1983). Severe blood disorders often occur during and after capture, so experimental studies are often carried out on animals which are not in a normal physiological condition.

Over the last fifteen years, great progress has been made in improving transporting conditions and in the maintenance of high water quality in sufficiently spacious pools. These factors enable something close to natural conditions to be achieved, and we can now manage to keep in captivity certain species of the big pelagic sharks (of the *Carcharhinus* group in particular).

THE SHARK'S BRAIN

We now know that sharks possess a great many attributes which were thought to be the exclusive apanage of the "higher vertebrates", what then must we make of their reputation as primitive bird-brained machines?

When the anatomists first became interested in this question, they worked on the two most commonly available species: the Spurdog or Spiny Dogfish (*Squalus acanthias*, a member of the most primitive group of sharks, the squalomorphs), and the Spotted Cat Shark or Lesser Spotted Dogfish (*Scyliorhinus canicula*, one of the most primitive members of the galeomorphs). They discovered small brains, which, added to preconceived ideas, perpetuated for a long time the myth that all the elasmobranchs had small-sized brains.

It was Northcutt who, in 1978, finally established the truth, showing that many elasmobranchs possess brains of a size comparable to that of many birds or mammals. It was subsequently possible to verify in the elasmobranchs, as in other animals, the relationship that exists between the level in evolution hierarchy and the magnitude of the ratio of brain weight to body size. So it is that the most highly evolved of the galeomorphs have a ratio 2 to 6 points higher than that of the most primitive of the squalomorphs.

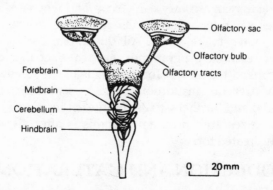

Brain of a Lemon Shark. The species with the highest brain/body ratio are the Dusky Shark and the hammerhead. Generally speaking, the active fast-swimming sharks have bigger and more complex brains than the slow benthic species.

The study of the brain of elasmobranchs might moreover provide an answer to one of the great mysteries of animal biology in general: that of the significance of this brain/body ratio outside the rank of evolution. Northcutt and Myrberg recently put forward the tempting hypothesis according to which the relationship between brain size and body size could reflect a process connected with metabolic activity. This hypothesis becomes most meritable when applied to the elasmobranchs, in which a disparity in activity exists which at first glance is as wide as the variations in the size of their brains. Thus, certain rays and skates (batoids), which stay hidden in the sand, have a very low ratio, while the galeomorphs generally show an intense activity. Many questions follow from the hypothesis and it will be interesting to examine the answers in the years to come: have the species which lay eggs a different metabolic level from those which have placentas, and is this reflected in different brain sizes? Is the locomotory activity of a species not proportional to the size of the brain in relation to that of the body? Positive answers would have deciding repercussions in the whole world of scientific biology.

These answers will be provided only by the fifteen or so scientists, all English-speaking, interested in shark behaviour, and even this very small number of researchers is declining every year. The shark is an animal that is particularly difficult to study in the wild, for it moves around a lot, rapidly, and very often solitarily or in small groups. Many interesting species are only abundant in geographically isolated regions and most are delicate, requiring very careful capture and transportation. Finally, the species most interesting to man are very dangerous, and when researchers have to enter the water in order to observe these monsters, strict safety measures have to be taken. All this explains the very long term nature of research programmes, and the fact that data banks are being supplied too slowly for the needs of the researchers and their sponsors. Certain institutions which set their sights only on commercial and business goals could earn respectability by granting aid to such research programmes, this is after all what they were originally created for.

REPRODUCTION AND MATURATION

REPRODUCTION

In this field, as in all others, sharks exhibit very particular characteristics and an exceptional adaptation. The reproductive method of other fish and of teleosts (bony fishes) is by comparison very archaic (thousands of eggs are laid in a seemingly haphazard

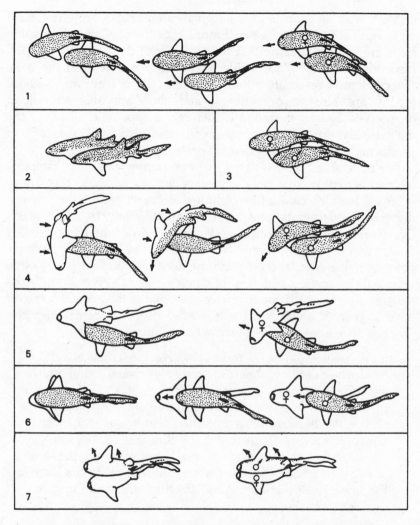

Postures associated with copulation in the Nurse Shark   (Klimley, 1980.)

1. Parallel swimming.

2. Biting of the pectoral fin (side view).

3. Biting of the pectoral fin (seen from above).

4. Revolving and about-face.

5. Female on her back.

6. Male mounting the female (intromission).

7. Male and female remain on their backs during insemination.

fashion, with an infinitesimal survival rate). Sharks are economical on energy, and produce a small number of foetuses and all possible steps are taken to ensure that the majority of them reach maturity. Insemination takes place inside the female, and at birth the baby sharks already resemble miniature adults. Their teeth and sensory organs are already operational, and they are able to defend themselves. Sharks' gestation is therefore comparable to that of the most highly evolved mammals, with the additional avantage of producing newborn young that are perfectly independent.

The male shark has no penis, but two pterygopodes (or claspers) which are visible. It is only necessary for one of these claspers to enter the female's cloaca (the common orifice shared by the urinary, digestive and genital tracts) for insemination to take place. Copulation has never been observed in the wild, the most detailed description is that given by Klimley (1980). The postures accompanying copulation explain the deep bite marks that are often seen on the bodies and fins of female sharks, even though the females' skin is clearly much thicker than that of the male. If you ever have a pair of sharkskin boots made, insist on female leather! Embryo development takes place in three different ways:

- In oviparous sharks, the most primitive, the female is fertilised inside the cloaca, the eggs are produced in small numbers, they are large in size, and they contain enough yolk to bring to maturity a newborn of good size. The eggs are enveloped in a shell which becomes hard on contact with water and which always has a very distinctive shape allowing it to become buried in the sand (such as the spiral shell of the sleeper shark), or to get attached to seaweed to shelter from predators (such as the "mermaid's purses" of dogfish which are often found on British beaches).

- Ovoviviparous sharks are the most numerous. The difference with oviparous sharks is that the eggs are incubated in the mother's abdomen and hatch inside her. Each embryo is attached to a large yolk sac by a cord, and feeds inside the mother but independently of her. Some ovoviviparous species exhibit a strange characteristic: "intra-uterine cannibalism". The first embryo that becomes functional begins to devour all the others in the mother's uterus, so that we witness the eventual birth of a single, fully formed baby shark. This macabre habit is in the end nothing other than a sophisticated method of natural selection, since it is the most successful (the most agile) embryo that devours all its defenceless brothers and sisters. It will come

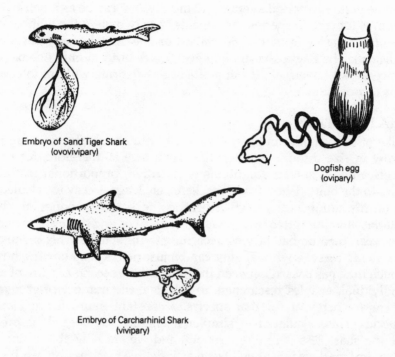

Embryo of Sand Tiger Shark
(ovovivipary)

Dogfish egg
(ovipary)

Embryo of Carcharhinid Shark
(vivipary)

The three methods of shark reproduction

as no surprise that this prenatal aggression is found in the young of the formidable Ragged-tooth and other large predatory sharks (see the directory at the end of the book).

The third method is the most highly developed: viviparity. Once the yolk has been consumed by the embryo, a placental connecting organ takes over, linking the foetus to the mother's system, as is seen in the most highly evolved animals. This method is found in the carcharhinid sharks, in the smoothhound (*Mustelus*) and in the hammerheads.

Whatever the method, the newborn shark requires no help from its mother, being immediately capable of surviving and living on its own. Besides its teeth and its sensory organs, which are functional straightaway, it possesses a large enough liver to enable it to fast for several weeks before finding its first prey. Some babies can reach 1.5 metres (the Basking Shark), the average being from 45 to 100 centimetres for the majority of species.

The gestation period averages 12 months, but can be as much as 22 months for certain species. As females have two uteri, the number of newborn varies from two to forty (for the Angel Shark *Squatina squatina*). The Ragged-tooth, however, gives birth to only one baby, since, as we have seen, the appetite of its offspring will not tolerate any competition.

MATURATION

Like other fish, and unlike many other animals, sharks continue to grow in size throughout their life. Their size is, therefore, for one species under similar conditions of nutrition, proportional to their age. In the bony fishes (trout, mackerel, etc.), age is easy to calculate, from the number of growth rings on the scales or bones. In the sharks, only the vertebrae harbour growth rings and there are several per year, the number varying according to the species. This is one of the many reasons why tagging can be justified, since knowing how much time has passed between the first and the second capture of an individual, enables researchers to calculate the number of grooves there are per year for that species, a constant figure for any one species. This calculation is simpler if tagging has been done on a young shark less than one year old, and we are now able to inject chemical markers which will leave an indelible mark easily identified when the animal is caught again (like tetracycline, which fixes on calcium). It has thus been possible to establish that all species do not grow at the same rate:

- 32 centimetres per year for Blue Sharks;

- 28 centimetres for the Mako;

- 4 centimetres for the Sandbar Shark (*Carcharhinus plumbeus*).

A very recent modern method of calculating age is a radiometric technique which measures the difference in concentration of a radionuclide which occurs naturally in the core of a vertebra at the time of birth and in the exterior growth ring. This method is set to replace all others in the near future.

SOCIAL BEHAVIOUR AND ECOLOGY OF SHARKS

The classic image of the shark as a solitary creature, hunting in its own domain for its own purposes only is not an inaccurate one for many of the big species, in particular the Tiger Shark, the Great White, the Basking Shark, etc. Many others, however, move about in shoals, and one of the most spectacular examples is without doubt that of the Scalloped Hammerhead (*Sphyrna lewini*). Entire populations of this species gather in the daytime around a number of islands in the Sea of Cortez. These great predators apparently possess mechanisms allowing them to reach deliberately specific areas within their hunting range, areas where they remain in a shoal during their non-feeding periods. In the Gulf of New California, at a depth of between 5 and 30 metres and a few hundred metres from the shore, I once saw an incredible procession of hundreds of hammerheads, varying in size from 2 to 4 metres, which were heading north at a speed of a few knots. The Mexican who was accompanying me on the dive confirmed that the same procession occurred every day at this period, at the same place, and was no doubt associated with a particular period of reproduction or food resource, the north of the gulf being well known for these periodic migrations of large pelagic species, whales as well as manta rays, sharks, etc. The spectacle was all the more impressive as our only possible refuge was the boat, thirty metres above us, as we were swimming over a trench several hundred metres deep. I feared that at any moment one of these monsters, less congenial than the rest, might suddenly appear out of the great blue deep and come at me, but nothing of the kind happened: they all passed by with indifference.

Some people think that such sharks might also meet up in very specific geographical zones at great depths. This is not in any way impossible, if we take into consideration how well developed their sensorial capacities are. As within any animal group, social interaction does occur in sharks, even if they come together only by chance and briefly. Here again, eye-witness accounts are few and far between, but instances have been reported of intraspecific association, sometimes revealing evidence of social-hierarchy systems with dominant and subordinate individuals. In these last cases, we may perhaps be seeing, as in a number of other species, a subtle example of anti-predator behaviour on the part of the subordinates, which protect themselves by associating with the bigger ones in this way. At least one clear case of intraspecific social hierarchy was observed in the bonnethead sharks, where it was noticed that females tend to avoid males whatever their size. This behaviour is no doubt due to the injuries that the males inflict on females when mating. This mating mainly takes place at night and proceeds in the same way for the various species. The male maintains a bite hold on one of his female's pectoral fins throughout the entire duration of copulation so as to keep the clasper in the cloaca pointing in the same direction. Fresh wounds on the backs of adult females suggest a real sexual harassment on the part of males before the gestation period (Clark, 1981). Perhaps this is why, in the Blue Shark, the skin of mature females is twice as thick as that of males of the same size.

We may wonder what other reasons there are for sharks to assemble like this in shoals. Although small sharks may be reducing the risk of individual predation, as soon as they reach adulthood the risk ceases to exist. Perhaps the answer lies in the fact that it is easier to find food when in a shoal, by pooling, for example, the increased detection capacities of the whole group. It is also known that the shark in a shoal often copies the behaviour of its congeners – the feeding frenzy being the most striking example. This is what we call "social facilitation" (Johnson, 1978).

One of the ways of understanding the activities of any animal is to find out what its role is in the ecology and the bioeconomy of the community of which it is a part. This is particularly so for predators, which can stabilise the dynamics of an ecosystem or cause an upward or a downward turn. Unfortunately, and we have seen the logistical reasons for this, observation of sharks is limited and has become compartmentalised into several fields, including that of ethology. Only a few studies have been made, usually commissioned by fisheries with commercial objectives. Much remains to be done, therefore, with the use of modern techniques: ultrasonic underwater telemetry, small one- or two-man submersibles (Nelson, 1981), underwater television, spectrophotography (Klimley, 1981), specially designed boats (Gruber, 1982) and even captive balloons.

One of the most useful avenues to explore would be to examine whether sharks communicate with each other and, if so, how. It would be very surprising if they did not when we size up all the exceptional sensorial capacities they possess for detecting their prey, and Myrberg has even wondered whether the shark communicates with its prey, which would be something of an accomplished refinement. One of the means of preventing the "shark threat" no doubt rests in interfering with or "jamming" its information system, or in sending out "counter-information", which could be acoustic or electromagnetic. In the light of the danger they pose to man, it is very important that we find out soon how the shark communicates.

THEORIES OF EVOLUTION AND THE SHARK

More than 500 million years ago, when all the continents were still only one gigantic mass (Gondwanaland) sharks and man shared a common ancestor, a certainty accounted for by successive theories of evolution. After Linnaeus and Lamarck, Darwin and the Neo-Darwinists arrived at a theory which was endorsed by the majority of scientists, that of mutation followed by selection by the natural environment. On a temporal and statistical level, I myself find it hard to see how this pattern of mutations and successive selections, even over millions of years, could have caused the living creature to change from a single-celled organism into a hypersophisticated mammal with highly complex organs and an extraordinary adaptation.

Among the many "schismatic" theories expressed is that of Motoo Kimura of Japan, who considers evolution to be the fruit of the accumulation of random micromutations rather than the result of the selection of the best of them. He considers that the majority of genetic mutations have no visible effect on living organisms, and therefore believes that they could only be brought about by chance.

Kimura looks for the cornerstone of his anti-establishment theory on the floor of the oceans, in the genes of the Port Jackson Shark. This fish's anatomy (see the directory) has remained the same for 300 million years, as a great number of fossils prove. Now, it so happens that this bottom-dwelling, very sedentary shark, possesses a globin (a blood protein used for fixing oxygen) which has been well studied, and which has changed as much as that of man, who is himself the fruit of continuous natural selection. Kimura deduces from this that "nature" cannot have exerted any natural selection on the globin, and therefore calls into question Darwin's theory. He calls his bombshell "the neutralist theory".

Whatever the future of this theory, sharks will play a part in attempts to prove it, since they are the animals that have evolved the least since they first appeared, as if chance had made them at the very start perfectly adapted to their environment.

7

1 001 Uses Of a Shark

THE SHARK AS POLICE-INFORMER

THERE ARE AT LEAST THREE recorded events in history, when sharks have been indispensable intermediaries in the detection of serious crimes. Each one of these stories is very well documented, supported by evidence in court, police reports and forensic reports.

On 3rd July 1799, the *Nancy*, a brig of 125 tons left Baltimore and sailed due south for forbidden waters. As an American boat, the *Nancy* was not allowed in the English Caribbean islands, but its owners had come to an arrangement to conceal their true identity. The boat steered as far as Curaçao, a Dutch colony of the Caribbean, where it obtained false papers indicating that it belonged to a Dutchman. Then it departed for the English islands, only to be intercepted by the English boat *HMS Sparrow*. The captain of the *Sparrow*, Huger Wylie, was not impressed by the Dutch papers, put an arresting crew on the *Nancy* and had it steer for Port Royal in Jamaica, where its case would be judged by the vice- admiralty.

Meanwhile, the crew of another English vessel, the *Ferret*, caught a shark in the jaws of which were stuck some papers. These papers were none other than those of the American boat *Nancy*, which the captain had tossed overboard, far out at sea.

As luck would have it, the captain of the *Ferret* invited Captain Wylie aboard, shortly after their miraculous catch. On examining the papers, Wylie immediately realised the deception. The "shark papers", as they were named, were produced in court, and the *Nancy* and her cargo were confiscated.

The shark's jaws, which measured 55 centimetres at their widest, are still on show, together with the papers, at the Jamaican Institute in Kingston as a warning to perjurers, because the truth will always out, even from the depths of the oceans, even from the jaws of a shark.

In 1914, the captain of the Norwegian boat *Gladstone* registered his ship in Costa Rica under the name of *Marina Quesada*, solely to provision German U-boats on the open sea from Virginia. In January 1915, once his missions were accomplished, the name *Gladstone* was painted back on the boat again and the Costa Rican flag was replaced by the Norwegian one. He then returned to Recife in Brazil, where the authorities asked to see his papers. The captain hurriedly put the papers of the *Marina Quesada* in a leather bag and threw them overboard.

The next day, the crew of the Brazilian warship next to the *Gladstone* killed a shark, and discovered in its stomach the papers of the *Marina Quesada*. Thus it was possible to prove one of the most flagrant violations of American neutrality by the German government. The captain of the *Gladstone* never knew how the papers had been recovered, the judges not daring to mention it.

One of the most extraordinary stories in forensic science since its inception involves sharks, and is known by the name of "the aquarium mystery" or "the shark-arm case".

On 17th April 1935, Albert Hobson set a shark line 2 kilometres off Coogee Beach near Sydney. The following morning, when he raised the line, he found that he had caught a small shark, but that not much of it was left, the rest having been devoured by what seemed to be a much larger shark. By a stroke of luck, the latter had got caught on the same line, and it turned out to be a superb Tiger Shark 4 metres in length. Hobson and his brother decided to tow their cumbersome passenger, which was still very much alive, to the Cogee Aquarium, which had never known such a fine resident.

From 18th April, the monster was put on show for the public, who came to visit in ever greater numbers. Fresh sea water was pumped into the aquarium daily and the prisoner accepted the mackerel fed to it. In fact for several days it swallowed everything on offer, but then, on 24th April, refused to eat anything at all, and became irritable.

The following day strange happenings occurred which were witnessed by a small gathering of fourteen people, among them Narcissus Young of the *Sydney Morning Herald*, who was later to give evidence in court: "The shark was moving around slowly, then suddenly speeded up, banging against the aquarium walls before diving to the bottom, where it swam in two or three irregular circles, before coming to a stop. I was three or four metres from the shark and clearly saw come out of its mouth a copious brown froth which smelled really foul. And then suddenly, up to the surface floated a bird, a rat, a load of muck, and then, in the middle of all that, a human arm with a piece of rope tied around it."

The police deposited the arm in the town mortuary, where the pathologist, together with a shark expert called Coppleson, examined it. It was a whole arm, including forearm and hand, clearly belonging to a very muscular man. The mystery might have appeared too difficult to unravel had there not been a very characteristic tattoo on the front of the forearm: two boxers in combat. In addition, a piece of rope 1.4 metres long was tied around the wrist by two half-hitch knots, and on the front of the arm, above the elbow, was a broad straight wound.

Coppleson thought that the arm could not have been bitten off, because it had been cleanly dislocated at shoulder level. The inner slope of the wound was not irregular, and there were no tooth marks on the cartilage of the humerus, which was still intact. The arm had obviously been removed with a knife. It had not been removed surgically since there were no clamp and forceps marks, so there was no need to examine the records of the latest amputations in the region's hospitals. These findings also eliminated the conjecture that the arm might have belonged to a soldier who had committed suicide by jumping from the top of a cliff and whose body had never been found. They also made enquiries in anatomy laboratories, but in vain.

Finally somebody suggested the seemingly extraordinary possibility that the arm could be linked to a murder, so the Sydney newspaper *Truth* published a description and a photo of the tattoo. The results were not long in coming, a man declared that the arm had to be that of his brother, James Smith, aged 45, a billiard- player and former boxing enthusiast. He identified the arm a few hours later.

During this time the shark became ill, and on 27th April it was killed and a post-mortem performed on it. A piece of another shark and a few fish bones were found in its stomach, but no other human remains.

Then began an eventful inquiry, unprecedented in the history of crime. The public prosecutor alleged that a certain number of men, including James Smith himself, Patrick Brady, a boatbuilder by the name of Reginald Holmes and several others, were mixed up in dishonest schemes, blackmail, death threats and attempted insurance fraud involving an enormous sum in compensation for the wilful wrecking of a yacht named *Pathfinder*. The scheme went wrong, and the quarrels that broke out between the conspirators must finally have ended in the death of Smith.

The story moved to Cronulla, a small seaside resort not far from Sydney. Here, in March 1935, Brady had rented a cottage called "Cored Joy" under an assumed name. Smith was last seen alive at six o'clock on the evening of 8th April, playing dominoes in a hotel in Cronulla with Brady and two of the hotel clients. Nothing more was heard of him until the discovery of his arm in the Cogee aquarium on 25th April.

Early on 9th April, Brady had paid the cottage rent in cash and taken a taxi to Sydney, where he had seen Holmes. The next day he went into two gloomy little stores out in the suburbs. In one he bought a mattress and in the other a 200 litre metal drum, and he took the lot back to "Cored Joy".

From all this the police deduced that Smith must have been killed during the night of 8th/9th April, and that his body must have been dismembered on the mattress and the pieces crammed into the drum. The drum had been taken out to sea in a boat, weighted with a breeze block and other heavy objects, and sunk in deep water. The police estimated that the whole body could not have been squeezed into the drum and that the arm must have been thrown overboard separately, after a sufficiently heavy object had been knotted around the bundle.

According to this theory, the shark would have got hold of the arm between the 8th and the 17th when it was caught, and the arm had remained a further eight days in the shark's stomach before being regurgitated on 25th April.

The police organised a detailed search in the waters off Port Hacking to recover the drum. They covered the whole region by plane and patrol boat and a diver was sent down, but found neither drum nor any body parts.

On 16th May, Patrick Brady was arrested and accused of Smith's murder. The next day, Holmes was interrogated by the police, and promised to give evidence to the examining magistrate.

Three days later, Holmes's outboard was seen forging along at top speed and apparently out of control off the port of Sydney. Holmes had collapsed in the cockpit. A police patrol launch and another boat with several men on board, including Holmes's brother, gave chase to the runaway outboard. When they finally caught up with it and got it under control, they found Holmes with a bullet in his forehead. He was taken straight to hospital, where a soft-tipped revolver bullet was successfully extracted from his wound, and was discharged fully recovered a few days later.

Seven days after the "accident", Holmes made his first swift statement to the police, but did not yet disclose any evidence. The night before his final statement was due, Holmes was discovered at the wheel of his car near Sydney bridge, with a bullet through his head. This time the shot had been a good one, Holmes would not be making any more statements.

Two men, one of them a very big shipowner in Sydney, were indicted for the murder of Holmes. Although they were tried not once but three times thus setting special precedents in case law, the court never managed to establish proof of their guilt. So one of the strangest cases in criminal history went unpunished, if not unresolved. Such are the drawbacks and the advantages of *habeas corpus*. John Patrick Brady died of natural causes in 1965, taking with him to the grave all the shady secrets surrounding the matter.

Another extraordinary story did not make the headlines, but I have been able to get hold of the most interesting pieces. This one took place a few years ago in South Africa.

On 7th December 1983, the nets set up by the Natal Shark Board off La Mercy (Zululand) brought in a Zambezi Shark 174 centimetres long and weighing 116 kilos. There was nothing at all unusual in the capture of this shark which is as common as it is dangerous on the coasts of South Africa. Like all the sharks captured by the NSB, it was quickly given a post-mortem, and in its stomach were found 2.5 kilos of human remains. A piece of fibula, a kneecap, some big pieces of flesh and in particular two intact feet (see the photos in the sealed section). These feet exhibited the peculiarity of having been perfectly dislocated, with the articular cartilages still intact and a complete covering of skin. From this it was deduced that at the time the shark had eaten the feet they could not have belonged to a live man trying to defend himself, but must have come from a corpse. The authorities reckoned that the feet must have belonged to a young Indian man, and inquiries were instigated and a notice placed in the local press. The very next day the detectives heard from a man who identified the feet as those of his son, who had been missing since 4th December. The witness signed written statements, and it was concluded that the young man must have drowned and the matter was considered closed.

At the Natal Shark Board, however, out of curiosity, they asked the man to bring in a pair of shoes that had belonged to his son. Instead of placing the shoes against the feet, they slipped a foot into each shoe, only to find to their amazement that the latter were much too big for these feet. The feet could not then have belonged to the man's son although he had identified them as such, and stuck by his statement. The NSB did not have the inquiry reopened, but it is certain that we have an unsolved detective mystery here. What mysterious motivation had driven the man to make this false statement, what sordid family drama had forced him to lie? The mystery is no nearer solution, despite the evidence provided, very much in spite of itself, by a shark with a taste for human flesh.

Hardly a year goes by without human remains being discovered by fishermen in sharks' stomachs. They are then automatically submitted for police investigation as soon as they are properly identified as such by the crew. Although presumably in reality only those with clear consciences will approach the police, others who, for example, have been fishing in waters that are out of bounds to them, probably get rid of their macabre finds without trying to investigate further. This was not, however, the case for the crew of the trawler *Ho Tai no. 12*, although they did happen to be fishing at the limits of Australian territorial waters in the Torres Strait in March 1988. When the net was raised at about midday, an enormous fish which threatened to tear everything to pieces had to be killed immediately with gaffs and harpoons. It turned out to be a shark, in the stomach of which were discovered the remains of a human skull, a thigh, a leg, some finger bones and a piece of underclothing "made in Taiwan", but not destined for export. It was deduced from this that the victim must have been an inhabitant of Taiwan, and, one month later, back at Kaohsiung, the deep-frozen remains were handed over to the police. The pathologists identified the remains as those of 27 year old Lo Ying-Chun, a Taiwanese fisherman on the *Lien Cheng Feu*, also based at Kaohsiung. On 21st January, when his trawler had been operating north of Australia, he had slipped on the freeboard and fallen in the water. The boat patrolled the area for several hours, but, as so often happens in these cases, when the man in the sea has neither whistle, nor mirror, nor flare on him to indicate his position, they did not spot him and he was reported as missing.

THE SHARK AS ALIBI

September 1985 north of Miami. Two boys leave on a fishing trip in the Atlantic on one of those big inboards specially designed for game fishing, with lines everywhere, sticking out on both sides like so many antennae.

A few hours later, one of the boys returned alone, in a state of collapse: "My friend was swimming around the boat, and suddenly a fin appeared.... The water was still red with blood for five to ten minutes after the shark had swallowed him."

The detectives recorded the statement without any comment, then one of them who knew a bit about these matters began to wonder: "That's odd; sharks generally attack from below, and it's rare to see a fin first. In particular, what is very surprising is there being blood in the water for five to ten minutes afterwards; blood disperses very quickly in water, especially on the open sea where the water is never calm."

The "victim" was discovered alive and well two weeks later, in Los Angeles, where he had gone after having stolen his ex-girlfriend's car, her jewellery and her credit cards, and having claimed on a life insurance of 200,000 dollars on himself.

2nd February 1989 Piombino. Luciano Costanzo, aged 47, goes diving off Tuscany accompanied by his son. They are two kilometres off the coast when to his horror Gian Luca sees his father disappear into the mouth of an "enormous white shark at least six metres long". The newspapers got hold of the story, ran headlines on the "assassin shark", and interviewed experts who certified that the highly dangerous Great White Shark did frequent the Mediterranean. The big hunt began, watched closely by an attentive television audience and the local authorities, who were worried about the coming tourist season. The search came to nothing.

The Livorno public prosecutor's office asked for a fresh report on the unfortunate diver's air bottles and on his lead-weighted belt "so as to determine the possibility of death being caused by an explosion". It was in fact thought that there could have been an accident while dynamite fishing, later falsified in order to claim on the life insurance.

There were no more signs of an explosion than there were of shark tooth marks. What they did discover, however, well arranged on the sea bed 27 metres down, were the lead belt, the bottles and the pair of flippers, all intact. As if the shark had carefully undressed its victim before swallowing him! No trace of the body, however, not even a part of it, was ever recovered. If we add here that shark attacks in the Adriatic can be counted on the fingers of one hand, this 6 metre Great White, which left no trace of itself and about which nothing was ever mentioned again, is decidedly most odd.

No wonder there were a lot of rumours circulating in the bistros of Piombino. Many believed Luciano Constanzo would be found in perfect health abroad, probably in France, leaving it to his son to collect the gold mine from his insurance premium. The shark therefore had to be given the benefit of the doubt and absolved.

Tales and superstitions concerning sharks came to the knowledge of the Old World at the same time as the great sailing ships were discovering exotic islands and strange lands where the shark was seen as both a god and a divine instrument. Civilised man was not long in discovering for himself certain additional uses, more functional than mystical. When the British established convict prisons in Tasmania at the beginning of the 19th century, they mounted guard on them with big dogs and armed men who regularly patrolled the camps. A number of prisoners, however, managed to escape from one of the camps situated at the tip of a peninsula. They slipped into the water at night, and swam away once the patrol had gone past, regaining dry land farther inland in the shelter of the mangroves. As a result the governor of the colony ordered that henceforth all refuse was to be thrown into the water all around the camp. Attracted by the daily promise of free meals, sharks rapidly concentrated at the site of the escape route. After a few episodes of screaming in the night, and after the prisoners had become familiar with the appetite of their new warders, escape attempts ceased.

Exactly the same method was used right up to the 1940s to prevent convicts from escaping from Devil's Island off French Guiana. On Royale Island, not far from Devil's Island, the casket can still be seen in which the guillotined bodies of convicts who had murdered a prisoner or a guard were placed. They only needed one coffin because the men were not buried in the ground. It was loaded onto a boat, which moved barely a hundred metres from the shore. The oarsmen then brought the coffin on deck, and the weighted body was committed not to the sea, but to the sharks. The boat had hardly departed for the shore before the waters became stained with the blood of the tortured victim.

Not only "aquatic guard-dog" or gravedigger, but also executioner, the shark's jaws can replace the hand of man to carry out the death sentence. Nobody will ever know how many slaves, dead, dying or alive, were delivered to the sharks in this way, but verbal reports at the time were numerous although discreet.

ECOLOGY AND THE FISHING INDUSTRY

Recent statistics compiled by the United Nations Food and Agriculture Organisation (FAO) reveal that the shark represents about 1% of the current market in fish, an enormous total amount. Having only in the last ten years or so become a regular part of the diet of countries such as the USA, the development of deep-freezing and preservation techniques, as well as the methods of culinary preparation, point to a major increase in shark consumption in the future.

On a more general level, and irrespective of the possible taste certain nations have for shark flesh, it is hard to see how the animal could escape the, often dramatically, increasing pressure of hyperintensive fishing, which is rendered unavoidable by a demand for protein proportional to a human population that has grown by 500% in 90 years. It is here in fact that the real world ecological problem lies, in this uncontrolled human overpopulation. If we had to find a single point in common between sharks and man, it might be that they have no true natural predators other than themselves.

The days are gone when man could believe that the planet's resources were limitless. Hardly has he discovered that all the sharks constitute major sources of protein when, ten years later, some species of shark are already on the road to extinction in certain regions of the world. Continental man passed from the Bronze Age to that of farming and stock-rearing several thousand years ago. Tragically, while he is already moving into the post-industrial age, man is still in the Bronze Age with regard to the oceans, continuing to draw on them limitlessly, with no real international legislation, and above all with no concern for breeding and rearing. Under the pretext no doubt that the marine element is not his own, he has not yet looked seriously overboard to realise the huge capacity for breeding and rearing that lies hidden beneath the surface of the marine biotope.

In the 1900s, the skate was very abundant throughout the Irish Sea. Among other fish species, this relative of the rays and sharks began to be fished there by various methods, and in particular by trawling. This unselective method of fishing has carried on right up to the present day, but skates are no longer brought up in the nets. None have been caught in the Irish Sea for over ten years. The disappearance of the common skate went virtually unnoticed and with no comment in the scientific literature. According to Brander, from calculations based on our knowledge of reproduction and feeding habits, we have for several years been capable of determining the threshold below which the reservoir of any given species will disappear. Again according to Brander, this threshold was crossed for the common skate only a few years ago. This skate is still present, in small numbers, in the waters bordering the Irish Sea to the north and south. It is, however, too late for it to return to this sea, unless all forms of fishing for all species of fish are suspended there for an indeterminate period. As this solution is currently unfeasible, the common skate represents for Brander the first indisputable example of the extermination of a species through uncontrolled industrial fishing. If the minimum non-restricting measures had been taken ten or so years ago in accordance with our increased knowledge as regards the management of fish populations, there is no doubt that the forecast would have been much better.

Compagno for his part cites the case of the School Shark or Tope (*Galeorhinus galeus*) of south of Australia. This species has been fished for since 1927, but under permanent monitoring. Whenever an alarming reduction was noted in catches of the shark, an investigation was set up and appropriate measures imposed. After sixty years of intelligent exploitation, the School Shark population shows no sign of becoming extinct.

As for those species thought to be dangerous, some might think industrial fishing would be the ideal definitive means of protection against them? Apart from the fact that this style of fishing could not be selective between dangerous species and others, we ought also to determine whether or not the permanent eradication of species, such as the Great White Shark, would lead to a serious imbalance in the animal world. Studies were made on the island of An Nuevo, 225 kilometres west of Baja California, by Le Boeuf and Ainley between 1980 and 1985, on the ecological relationships between pinnipeds, seals and sea elephants, on the one hand, which are very abundant on the island, and on the other the Great White Sharks, which are very numerous in the surrounding waters. As early as 1885, Townsend had noticed that 25% of female sea elephants bore deep scars from shark teeth. Few young, on the other hand, displayed such after-effects, leading him to think that they do not survive bites owing to their small size. Again on the same island, Le Boeuf established that, out of nine wounded females, only one was able to rear its young successfully. It is conceivable that, the moment the seal or porpoise population becomes very abundant, the Great White Sharks change over a few years from a mainly piscivorous diet to a diet centred on these mammals. They then become geographically tied to this region, and the food they take allows them both to select out the weakest elements among their new prey and at the same time to control its overpopulation. If the Great Whites in this region did not exist, or did not change their feeding habits at the right time, there would be overpopulation among the seals, and a considerable reduction in the numbers of fish. Here, then, there is a classic ecological balance in which man does not play a part since the An Nuevo islands are uninhabited. The time when this balance may be questioned is when seal colonies, moving closer to urban sites, may be deemed responsible for an increasing presence of Great Whites, and potentially a greater number of attacks. This problem was a topical one in 1990 in some parts of the globe, and in California in particular.

This correlation between dangerous sharks, the weight of commercial fishing and ecology can be illustrated again in an interesting fashion by the current situation of two well-known sharks in Lake Nicaragua. These are the Bull, Zambezi or Nicaragua Shark or *Carcharhinus leucas* and the Nicaraguan Sawshark (another elasmobranch). Only the first is a real danger to man, the second has

Map of Lake Nicaragua, kingdom of the Bull Shark.

never been known to attack. The two fish habitually enter shallow fresh waters, and in particular have established themselves in the huge Lake Nicaragua in Central America. This lake flows into the Caribbean Sea via the San Juan river, regularly followed in both directions by the two fish in accordance with their breeding cycles. We have already discussed the impact that the Bull Shark has always had on the populations of American Indians around the lake (see *Devil sharks and god sharks*) on account of its very old habit of attacking man.

Thomas B. Thorson (University of Nebraska) made a very long study of these two fish from 1960 to 1982, as part of an official programme which set out to tag as many fish as possible so as to see what their migratory cycles, their growth rates, etc. were. In all 3500 Bull Sharks and 377 sawsharks were tagged. It was through this operation that it was possible to confirm that regular and rapid migrations took place between the lake and the Caribbean Sea. The Bull Shark has its greatest population density in the lower part of the river and at its mouth, and breeds in the adjoining brackish waters, but not in the lake. The higher upriver one goes, the lower the number of sharks. By contrast, the sawshark is very abundant in the lake, where it breeds, and its concentration diminishes as one moves downriver. In the case of the Bull Shark, the population in the lake is maintained at a constant level by recruitment from the coastal population, while for the sawshark recruitment is achieved almost entirely through new individuals being born in the lake.

Studies in 1960 found the Bull Shark population to have been in slight decline for a few decades, although not in any significant way, and Thorson had no difficulty in finding sharks to tag, not only at San Carlos in Nicaragua but also at the point where the lake flows into the San Juan, and at the Rio Colorado in Costa Rica, where the main arm of the Rio empties into the Caribbean Sea. However, the decline became more distinct in the 1970s when three or four entrepreneurs from Rio Colorado started buying sharks right through the year for their flesh, fins and sometimes their skins.

Several thousand sharks were caught every year, and this could not fail to have repercussions on the population at the river mouth. In 1982, Thorson calculated that only 10–20% of the sharks entering the mouth got as far as the lake. Therefore, the population in the lake could only decrease in parallel with that at the mouth. His tagging figures at San Carlos confirmed his fears: in 1965, a few hours fishing in the lake brought in several sharks; in 1975, the new government made a decision to do something about the decreasing numbers by banning fishing for profit for two years; but good intentions were not followed by good actions, and political developments in the country have prevented further measures. I visited Barra del Colorado myself in 1984 to fish for tarpon. While I was there, I asked a Costa Rican to tell me a good place for shark-baiting; he told me to stick to tarpons as there were many more of them, so I let it go.

Lake Nicaragua was indisputably the best place in the entire world for sawsharks up to 1970, for no commercial fishing had ever been organised there. In that year, Thorson caught and tagged 252 sawsharks off San Carlos in 43 days. The record was 23 in a single day. But then commercial fishing started. At first it was just a small business in the northwest of the lake, at Granada, for catching and freezing the fish. However, within a few years there was organised fishing with nets over the whole lake. In 1974, Thorson caught only one fish in five days, 30 times fewer than four years earlier. In 1976-1977, he had a team of three men at Barra del Colorado, and three at San Carlos. During the whole of that year, these men caught 11 sawsharks, or one per month. Thorson conveyed his alarm to Managua, but nothing was done – perhaps because the owner of the largest sawshark-fishing company was none other than a government minister who couldn't care less about the extinction of the species. A case, no doubt unique, of a single man being capable of exterminating a species of fish within a few years. In 1980, the new government decided that the catch would be limited to 113 tonnes per year, and fishing banned during the four breeding months; it later instituted a two-year moratorium banning all fishing for sawsharks and other sharks. According to Thorson, at such a rate it would still take ten to twenty years for the population to recover to an acceptable level.

These few examples show not only how the noble line of the lords of the sea is not safe from the risk of extinction, but also the speed with which excessive fishing pressure can bring about this extinction. This is due to a number of factors:

- the relatively slow growth of these sharks, and the long period necessary to reach reproductive age;

- the fact that they have relatively few offspring;

- their long gestation period (one of the longest in the animal kingdom).

We now have sufficient scientific knowledge to be able to provide effective advice to fisheries administrators in order for them to find a balance between real needs, profit margins, ecology, and also, for the relevant regions, the "shark danger".

We should never again have to see a repeat of what happened in November 1986 on the Californian coast. The Thresher or Fox Shark (*Alopias vulpinus*) was very widespread up to this time off Los Angeles, and was always regularly fished here as indeed it was in other parts of the globe, especially Europe (see the map at the end of the book). This species moves north in summer, and heads back towards the south of the west coast of the United States in autumn. In that November, some fishermen who had been using a purse net were travelling in a spotter plane, when they saw an enormous shoal of Thresher Sharks heading southwards following the coastline. Ten or so fishing boats immediately converged on the area, and spent two whole weeks catching hundreds of the poor shark, exterminating the entire shoal. In the aftermath of this veritable "genocide", the authorities were very late in asking themselves whether the species could ever be restored in this region of the globe.

Beulah Davis told me a story of how an American senator, having heard that some male sharks lived in the north of the Pacific while the females lived right down in the south, decided to put them all together to create a profitable shark fishing industry. This only goes to prove that an American can be a senator and at the same time ridiculously ignorant of the most elementary rules of nature.

THE SHARK AS A GAME FISH

The technical advice in this section is taken largely from Rob Hughes, the great Florida fisherman who has more than 8,000 sharks to his name, including 2,000 caught with rod and line.

I have mentioned elsewhere the importance of industrial fishing (300,000 tonnes in 1984), and the danger it represents for a species which reproduces very slowly. I have also detailed the different techniques of net fishing carried out near coasts for the prevention of attacks, and emphasised the remarkable effectiveness of this method both in Australia and in South Africa. Sport-fishing of the shark is relatively recent so far as its approval by the very responsible International Game Fish Association is concerned. Only six sharks are recognised as truly sport fish: the Great White, the Tiger, the Mako, the Porbeagle, the Blue and the Thresher. These fish can be caught at any time of day or night from a bridge, a pier or a boat, near the shore, offshore or at open sea.

If in a boat, you must make certain that it is big enough to be stable in high seas, it should be at least 5 metres long, and that its freeboards are high enough so that it does not become flooded at the slightest listing, and that it has a wide enough beam so as not to capsize with the first big catch. Furthermore, the hull should be solid

Complete trace for shark-fishing. Pieces of fish or mammals, or even a live fish, can be attached to the hooks.

enough not to split at the least blow from a discontented shark's snout (see *Attacks on boats*).

The length of your fishing rod depends on the method of fishing you have chosen. From a pier, the rod must be shorter (approximately 2.7 metres) and stronger so as to give better control. From a boat, it can be longer and more flexible, which will allow the sharp tugs by the shark to be absorbed. Not being limited by space, one can follow the shark that flees merely by starting the engine. The best rod is made of solid fibreglass (hollow ones can break) and measures 2.7 to 3.3 metres. The guides should be on rollers, as this is the only way not to scratch them and risk breaking the line.

The choice of reel again depends on your method of fishing. From the shore, a greater length of line is needed, and you should opt for a 124/0 to 16/0 with 800 to 1,000 metres of line tested to 130 lbs (a British measure, used throughout the world). From a boat, a 10/0 reel with 600 metres of 80 lb line is required. Do not forget to secure the wing of the reel to the boat, this will avoid the risk of seeing the line go overboard. The reel can be attached to your body harness to make it easier to negotiate the shark's fleeing movements and sudden turns. Do not skimp on the "padding" between the harness and the butt of the rod: you will understand why when you have to put up with this butt pressing into the pit of your stomach for three hours.

The rod handle must be covered with a material that is non-slip, even when soaked with water or coated with fish mucus. Cork or rubber generally suffice, especially if the diameter has been adapted to the size of your hand. Some fishermen make themselves a handle based on resin moulded on their hand, which is all very well as long as the rod does not turn.

The line is the third determinant item in the success of a shark-fishing trip. You must never go below 80 lbs, and, if you are really going for very big fish, you will have to opt for 130 lbs; you may perhaps cause your neighbours to smile at your optimism, but the smiles will disappear if you hook a 1,000 lb Tiger Shark.

For the line trace you have a choice between steel monofilament or plaited wire. Each has its advantages. Monofilament will pass between the shark's teeth and cannot be cut in two, but it can get twisted or break with pulling. Plaited wire is too big to pass between the teeth and runs the risk of being cut, but in return it will not twist and will not break.

The choice of line trace also involves its length as well as its strength. We know that the shark's skin is extremely rasping, and no non-metallic line can stand up to it for long when the shark winds it around its body as its struggles; the projecting edges of the fins very quickly achieve the same result. It is therefore essential to choose a sufficiently long metal wire: 3 metres (for boat fishing) to 7 metres (for fishing from shore, with risk of rocks). A good size is the 1/16 stainless steel (it is in fact twisted and not plaited).

The hook must be made from a single piece of steel, and not turned over on itself and welded, in order to avoid it opening up when pulled on. Its size can vary but the 120 is a good intermediate size. These eye hooks will get rusty in the area of the "eye" if they are not cleaned, but they have much better resistance than ordinary hooks, even welded ones. The line trace should be mounted as per the drawing above with bushes press-fitted.

Before casting the line, there is still the choice of bait to be made. Shark bait has to be both oily and bloody. These two cardinal qualities ensure a good diffusion in the water and detection at long distance. In Florida, for example, the grey mullet, the bonito, the stingray and the bullhead are considered good baits, with the mullet and the bonito topping the list. You must cut the bait into pieces of between half and one kilo. Some fishermen think that "the bigger the

bait, the bigger the catch will be". However, a small bait can be rapidly swallowed, while a large bait may simply be held in the mouth while the shark does a half- turn, and thus be torn off when the line tightens, before the hook has even had a chance to embed itself.

Many fishermen use the assembly shown below with a live fish, which has the advantage of acting as a lure corresponding to the shark's usual prey and of emitting vibrations, which we know the shark to be very susceptible to.

Sometimes the mounted hook is not enough, and baiting is the only method of attracting sharks to the boat. In that case you need to put whole fish into a meat mincer and hurl handfuls of this mixture over the stern. This method is not always considered sporting, especially in some parts of the world where there are lots of sharks. Some fishermen compare it to shooting a deer that comes to eat corn which has been spread on the ground by the hunter beforehand. What you must never do under any circumstances is lure with bait from the shore, for this would not only attract the ingenuous sharks you are seeking, but also the big predators, which are not averse to taking fishermen (a number have been seized in this way in less than a metre of water).

Bob Hughes recommends using a "clip rope" in order to avoid losing the shark that you have managed to bring in up to the side of the boat – loss at this stage is a common problem. I recommend the

To line trace

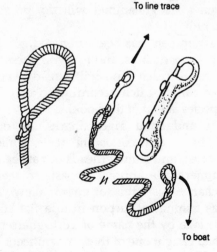

To boat

The "clip rope", a safety measure

use of a nylon rope of 3/8 to 1/2, a few metres in length (according to the design of the boat) and comprising two loops at each end. One of the loops is attached to the steel trace by means of the fast-grip clip, and the other to any cleat or sturdy mooring point of the boat. It is then possible to get rid of the rod, or even to cut the line, so that the only thing you have to deal with is hauling the animal aboard. It must be noted that even an apparently tired and listless fish always keeps a spectacular reserve of energy for struggling as soon as it is taken out of its element. All those who have lost one or more fingers, sliced off by a 130 lb line when a shark returned to the deep, are well aware of the problem. This nylon rope can itself be attached to a "shock rope", which is in fact made of stout india rubber and allows shocks to be absorbed.

Once the shark is out of the water, it must still be gaffed if it is to be hauled aboard. It may be gaffed anywhere, but it is best to aim as near to the the gills as possible. The ordinary gaffing spear is dangerous and a number of fishermen have been knocked senseless and died while they were trying to finish off their catch. The "flying spear", especially one with four spikes, is much safer. If you fish from shore, the ideal grappling iron is the one used for scaling walls, with three or four barbed spikes.

For finishing off the animal, the "bangstick", which was described in the section on preventative means, is ideal, since it fires a no. 12 cartridge into the head of the shark at point-blank range. The baseball bat is more "sporting" and very effective if you know how to use it, or even a sledgehammer will do, provided you do not smash the boat's hull.

Once all this equipment has been prepared, all that remains is to catch the shark of your dreams, and to try to guess what species it is while it is still out of reach. You will already know what type of shark you can or cannot catch according to your location (see the maps for each species at end of the book).

Lemon Sharks and Sand Tiger Sharks are often fished near beaches, Nurse Sharks near corals and reefs, while Tigers are met with in deeper waters and hammerheads out at sea. If it is the tarpon season, the hammerheads approach coasts to seek out this select prey. I had the chance to see this for myself a few years ago in Costa Rica when I was fishing for tarpon from a flat-bottomed boat 200 metres from a village by the name of Tortuguero on the Caribbean coast. A friend had caught one of these magnificent tarpon, but I was less fortunate and beginning to feel thoroughly fed up with this static and futile fishing. Suddenly I caught sight of a spectacular wake from something heading towards my walnut hull, the gunnel of which was just 20 centimetres above the water, and there below us was a magnificent hammerhead shark. The local fisherman who was with me told me that there was no point in carrying on fishing for tarpon with a monster like that in the waters, and I did not argue.

How the shark behaves on the hook will also help you to work out what it is. The Nurse Shark puts up a fierce struggle and twists and turns when it is close to the boat; the Lemon Shark fights only for 150 metres; the hammerhead pulls hard and does not give up easily; the Mako leaps out of the water and fights on for a long time.

The time of year is also a guiding factor according to the fishing area. For example, in Florida, the winter period from October to April is associated with the presence of Brown, Dusky, Bull and Tiger Sharks, whereas the summer from May to September brings Lemon, Nurse and hammerhead sharks, and again Bull and Tiger Sharks. Finally, do not forget that the only safe way to remove a hook from a shark's mouth, is to wait until the next day.

A love of game fishing, allied to a frenzied fascination for sharks, can lead to somewhat bizarre excesses. A few years ago, a very youthful retired gentleman from the Caux region on the north coast of France undertook to implement his lifelong dream: to fish for sharks in the English Channel.

On his very first trip out from the little port of. Saint-Valéry-en-Caux, near Dieppe, our man captured a 50 kilo Porbeagle. This was all that was needed for him to be overcome with euphoria, and as a result he founded an organisation, unique in France: the "Saint-Valéry Shark Club".

The club was set up in the old customs office, which had a weighing machine just in the right place, right on the quay where future catches would be unloaded. A superb coat of arms soon crowned the entrance to the building, and the only thing left was to organise a competition to give the new club its aristocratic pedigree.

The first competition drew fishermen from the whole of Europe, and on a fine morning in September all the teams took to sea for six hours of adventure. A number of sponsors had offered five or six cups for the event; the Altea Hotel accommodated the biggest champions; the regional newspaper, the *Paris-Normandie*, ran the front-page headline "Jaws at Saint-Valéry-en-Caux"; the customs scales had been checked – in short everything was absolutely ready for an international competition worthy of those which regularly take place in the tropics. The only restriction laid down by the adjudicating committee of the event was a total ban on baiting with coagulated blood, in order to avoid excessive catches.

When the time for the prize-giving came, at the end of a very full day, a certain perplexity could be seen on all faces. The first prize was in fact rewarded for the only catch of the competition: a dogfish weighing 2.5 kilos, not even ratifiable by the international authorities since its weight was below 5 kilos. Lots were therefore drawn among the friendly participants for the other cups, and they parted empty-handed but happy all the same. In Maupassant country (Normandy), cider and calvados always make up for the greatest disappointments.

SHARK CUISINE

I have not yet dared talk about gastronomy in connection with the shark for I am French and the French are always fastidious on this subject, but it should be known that the shark is a highly sought-after food in Asia, where at least twenty-five classic ways of preparing it exist in Japan alone. In the Yucatan, fillet of young shark is one of the dishes most in demand by the descendants of the Aztecs, and dried salted shark is sold everywhere in Mexico. In the USA, the frozen fillets of white fish that are found in supermarkets under the name of "cod" or "swordfish" are very often actually shark. Nearer home, the popular British fish and chips very often consist of shark rather than true bony fish, and in particular of Porbeagle, which is very abundant, especially in the Irish Sea.

So it is not necessary to be in a life-and-death situation to feed on shark, and you may perhaps be happy to know a few basic preparation methods for the day when you wish to justify your catching a lovely shark to your ecologist admirers.

First of all you should know that not all sharks taste the same, and that the following six species are considered to be the tastiest: the Great White Shark, the Mako, the Porbeagle, the two kinds of thresher and the dogfish; after this come the Small Dusky and the hammerhead sharks. The flesh of the huge Basking Shark for instance is too flabby to be eaten. No shark is poisonous, however their livers are to be avoided, often being much too rich in vitamin A, although unlike those of the swordfish, they do not contain high concentrations of mercury.

Another elementary principle is to know that shark's flesh very quickly takes on an ammoniacal taste, and so it must be prepared and chilled as soon as possible after the animal is captured. It must be cleaned and gutted like any fish, the skin removed along with the dark flesh that is found just beneath it, and the fillets not cut too thick. The latter should be coated with flour and placed in the refrigerator for at least one night, again to avoid the taste of ammonia.

If you are on an island, an uninhabited coast, or an isolated boat, shark meat is ideal for smoking, but you will need to be patient: six to eight hours in a special oven at 115°C, 25 to 30 hours in the open, for fillets 3 centimetres thick which you will have salted and basted in margarine beforehand.

As for recipe details, I prefer to refer readers to the specialist works, although I will point out that this fish cannot be steamed or braised like many bony fish with a very delicate flavour. You should also be very generous with the green lemon, the onion, the garlic, the salt, the Cayenne pepper or chili powder, the paprika, the olive oil, etc.

The fins soaked in brine for two days and then dried are highly prized as raw material in oriental markets. When boiled, their cartilage releases a smooth gelatin which gives body and taste to soups. Note that for this preparation only the dorsal and pectoral fins are used.

And then, if you do not have the courage to prepare shark meat for yourself, you can always cut it into into big pieces with the skin still on, as it makes the best possible bait for crabs. You can also feed it to your cat or dog; it is one of the main food sources for domestic animals and livestock in many countries.

If dietetics will convince you, you will be happy to hear that shark meat contains 7% protein, very little cholesterol and is very nourishing.

THE SHARK MEMENTO

The tourist industry has considerably increased the demand for all those little objects and other knick-knacks in poor taste which some package-tour travellers like to deck themselves out in. The shark is a prime target for these modern-day Tatarians who are mad about shark teeth mounted on pendants, open jaws for their walls, bracelets made out of vertebrae, or in a word about all those trophies captured by others which make you look like a man when you get back home. Even the shark's eyes have found a place on the market. After having boiled them for an hour or so (preferably outside rather than indoors), you have only to cut them open to be able to remove that delightfully-shaped, little lens, now hardened, which is known as the crystalline lens. This lens can be dried, pierced, mounted and even renamed by a jeweller who will sell you a "shark pearl". These pearls are a real hit with American tourists. Another brainwave in the same style involves threading a few dozen cartilaginous vertebrae, which have been bleached beforehand in chlorinated water, on to a metal rod, interlocking the lot, fitting a metal knob and selling the object as... a walking stick! I would not advise you to rely on it as support for a broken a leg, but, if you ever run out of conversation, you will have a magnificent talking point. I must add here that, besides these slapdash pieces of trash for "suckers", I have recently come across a 19th century walking cane with a finish, a sturdiness and a patina that are all magnificent. The friend who sports this object uses it to ease a painful back complaint, and the admiral who presented it to him as a gift on behalf of the "Royale" in this case chose, beyond any doubt, an object worthy of his rank.

SHARKSKIN

The shark's hide is one of the main trophies of shark fishing, for it lasts an extremely long time if it is prepared correctly, namely if it is well flayed and then washed, cured, dried, etc. The dried skin is known as "shagreen" because of its similarity to the untanned granulate hide of the back and hindquarters of a horse (this leather also being known as shagreen).

The most sought-after skin which fetches the highest price in the leather industry, is that of the Tiger Shark. Hammerhead skin is among the cheapest. The denticles are removed before the skin is tanned to make a handsome and durable leather which is used for quality shoes and cowboy trousers. It is even more elastic than cowhide or pigskin and much sturdier (150 times more resistant than that of bovid leather).

The Ocean Hide and Leather Company of New Jersey was the first and the biggest producer of shark leather, but is now running into supply difficulties and having problems satisfying demand since the commercial shark fisheries are having problems. Not enough is known about the reproductive rates of sharks, but we do know that the replacement rate around the fishing areas is too low, equally in the United States, Mexico or Cuba as in Australia or England. It may seem that the solution lies in ocean fishing on the high seas, which is what the Japanese already do.

The Japanese are the world's leading shark-catchers and have a long tradition involving sharks. This is why samurai sabres have always had hilts covered with Angel Shark hide, the roughness of which prevents the sword from slipping in the hand.

This skin was also used for a very long time as sandpaper before it was manufactured by industrial methods. To prove this for yourself, you need only pass your hand over the skin of a shark in a tail-to-head direction, it is extremely abrasive while going from head to tail it is amazingly soft. In Sumatra, the skin of the Angel Shark is used to make drum skins. Up to the beginning of this century, it was fashionable to cover various personal items with polished shagreen: spectacle cases, jewellery boxes, book bindings, etc.

A final type of highly prized leather, the "Boroso leather", comes from the processing of Moroccan sharks. The dermic denticles are not removed, but polished so as to give the leather a texture that is both aesthetically pleasing and very tough.

THE SHARK IN PHARMACY AND IN SURGERY

In the 1930s the liver of sharks was discovered to contain oils and vitamins in great amounts. A veritable explosion of shark fisheries followed, notably in California, where after a few years a steep decline in the population of the "Soup Fin Shark" was recorded.

When the benefits of vitamin A on the night vision of wartime fighter pilots were discovered, the numbers of sharks taken became uncontrolled and there was what was called the vitamin A "gray gold rush". This rush abruptly ceased when a more economical means was discovered of manufacturing certain vitamins synthetically. The fisheries had to close, and the population of the Soup Fin Shark (*Galeorhinus zyopterus*) returned to normal in only two years.

The squalene contained in the liver of numerous sharks is a much sought-after oil in cosmetics, and is what lies behind the interest of some fishermen in the enormous Basking Shark, whose liver is in proportion to its size. Shark liver also provides basic materials for manufacturing lubricants and paint bases.

In the United States, corneas of elasmobranchs (fish with cartilaginous tissues), and more particularly those of sharks, have been successfully used for grafting onto the human eye. They have the advantage that they do not dilate when the salt content in the surrounding environment varies, unlike those of the bony fish (teleosts).

The cartilage of sharks is used in the treatment of serious burns, and some liver oils may be anticoagulants and may reduce cholesterol levels.

THE SHARK AND THE CINEMA

I have attempted to explain in another chapter the reasons for man's fascination for the shark, and it is significant that the greatest success story in the history of the cinema up to the 1980s was attributable to this terror of the seas. Steven Spielberg allowed no technical obstacle to stand in his way in making his Great White Shark true to life. Three robot models were perfected by Robert Mattey, former head of special effects at the Walt Disney studios. Made of plastic, they weighed 1,500 kilos, measured 8 metres in length, cost 150,000 dollars each, and were called Bruce. It has to be said that the bosses at Universal Studios in Los Angeles naively imagined that they could film real Great Whites on location in their natural element! All the powers of persuasion of an Australian man-and-wife team specialising in undersea filming (Ron and Valerie Taylor) were needed to get the go-ahead for the models to be made instead.

Is the film credible? I discussed this with Professor Compagno in 1989, when he was director of the Grahamstown Institute of Ichthyology in South Africa. He had been the film's technical adviser for the most spectacular sequences, and he confirmed that, taken individually, all the situations are plausible. From the size of the shark to the targets of its attack, from the size or the power of its jaws to its resistance to harpooning, from its method of attacking from below in deep water to the surface attack on the beaches, all the events depicted are possible and have already occurred on many an occasion at various places around the world. There is hardly anything that can be criticised apart from the rather excessive accumulation of attacks, and above all the animal's anthropomorphic determination to have the hide of its three pursuers by no matter what means. This film was followed by several others which claimed to be in the same vein, but in which the improbability of the situations vied with the absurdity of the scenarios. Professor Compagno was not consulted on these purely commercial productions.

Another feature film ought also be mentioned with regard to sharks and more particularly to Great Whites: this is *Blue Water, White Death*. Made earlier than Spielberg's film, it did not have the same international impact for it was much less commercial, but the heros were real sharks, not models.

THE SHARK VIRILITY TEST

The Blue Shark is a cosmopolitan fish and a dangerous one, capable of reaching 3.8 metres in length and credited with a number of attacks on people and boats.

One particular "sport" has developed in certain diving clubs in south California which involves swimming among Blue Sharks. To be certain of their presence, they are first of all lured with blood and pieces of fish thrown around the boat. To date, this "test of manhood" reserved for male divers has not yet claimed any lives, but the day a big Blue Shark or a Tiger Shark comes along to liven up the initiation ceremony, things may change.

In fact the Blue Shark is not very aggressive when mixing with

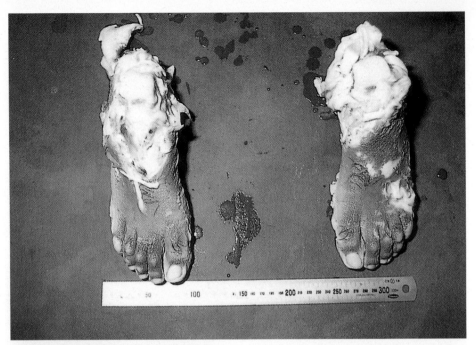

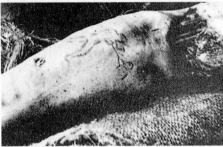

The "shark arm case": without doubt the most incredible case in forensic medicine worldwide. *(D.R.)*

(OPPOSITE) In 1983, the Natal Shark Board brought up in its nets this formidable Bull Shark (better known as the Zambezi Shark, *Carcharhinus leucas*), weighing 116 kg and 1.74 m in length. During the autopsy, its stomach was found to contain 2.3 kg of human flesh such as these dismembered feet above. The start of a detective riddle … *(NSB/X.M.)*

All sorts of things are found in shark's stomachs, such as this human skull with the facial bones torn away. It is most unusual for attacks to affect the face and head. *(NSB/X.M.)*

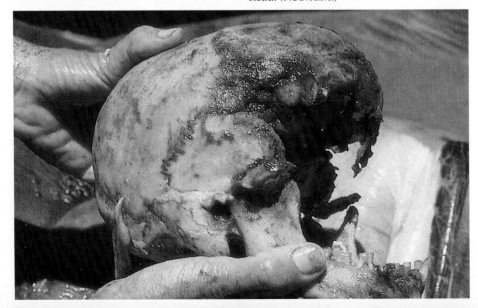

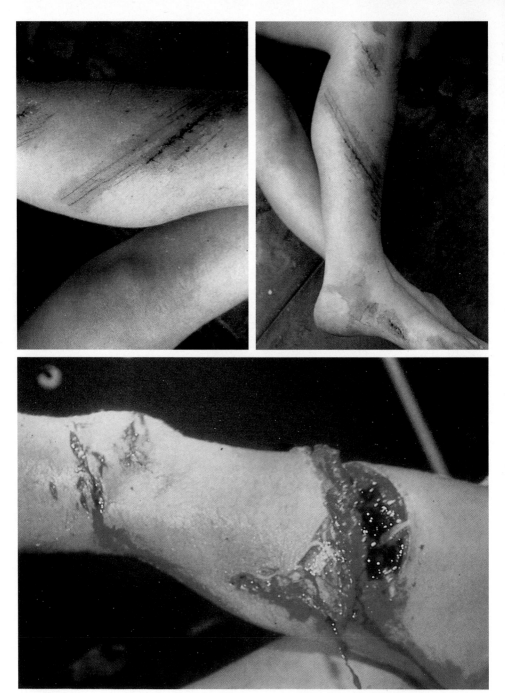

The injuries inflicted by a shark can be minor when they are due to chafing by its very rough skin (TOP). They become more serious as soon as the shark makes use of its formidable jaw (ABOVE). *(NSB/X.M.)*

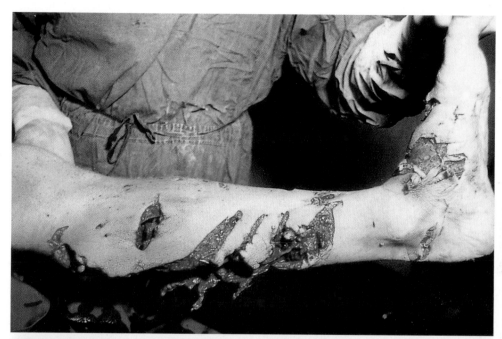

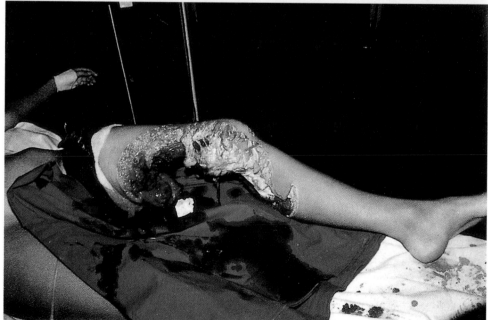

Injuries can be very serious, as in the case of Bruce Eldridge (TOP), and very soon prove fatal in the absence of immediate assistance (ABOVE). This assistance involves simple first-aid (tourniquets, vascular compression points. *(NSB and NSB/X.M.)*

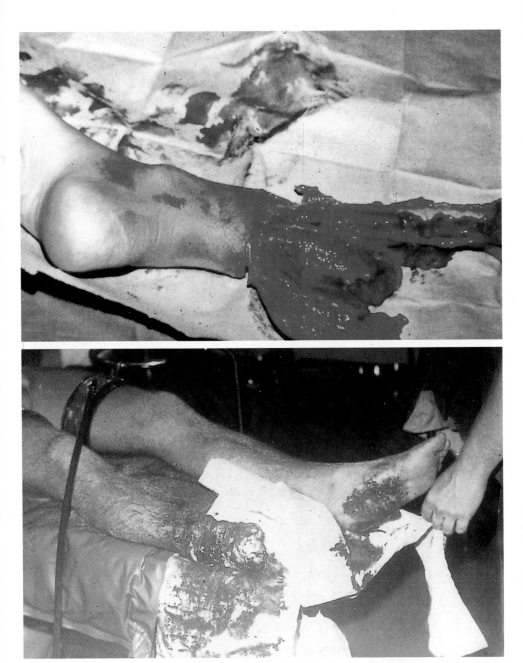

The injuries often necessitate amputation below the knee. Saving limbs is impossible when the bite has been motivated by hunger, as the flesh has been torn away and the internal edges of the wound are too shredded. (NSB/X.M.)

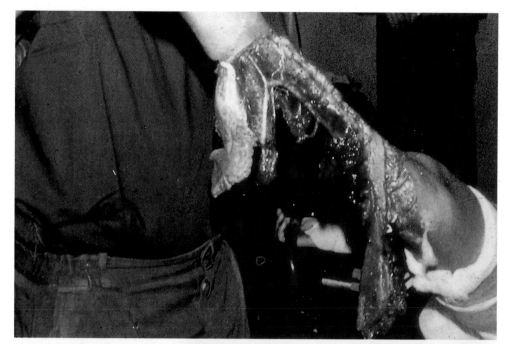

Injuries regularly necessitate amputation very high up towards the base of the limb. (NSB/X.M.)

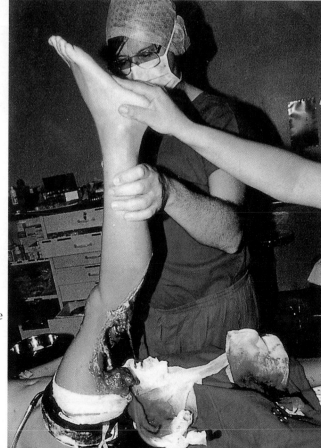

In this case the victim did not die of her injuries despite the femoral vessels having been torn away. In such instances the injured person may die from venal and not arterial haemorrhage, for the femoral artery goes into spasm in the vast majority of cases. A simple compression can therefore arrest the venal bleeding. (NSB/X.M.)

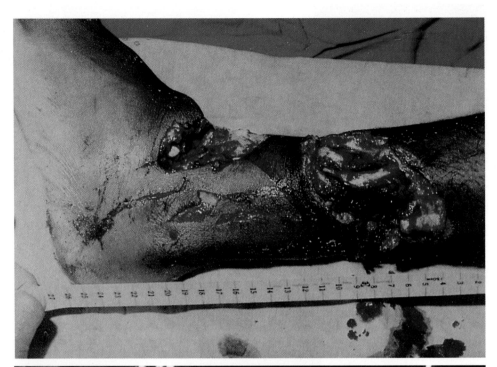

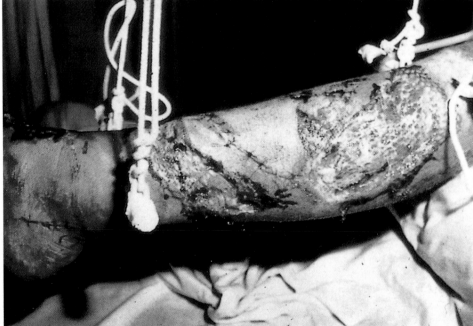

Many victims recover from their injuries, despite their often horrendous wounds and the danger of infection from the numerous germs contained in sharks' teeth (the shark is omnivorous and may feed on putrefying carrion). *(NSB/X.M.)*

Henri Bource, diving again after the attack that cost him his leg in 1964, holds open the mouth of a great white shark – the species that almost ended his life. He claims that he is not bitter about the incident, saying simply that: 'They [sharks] do what nature intended them to do...'

Attack does not inevitably signal the end of a passion. Henri Bource was attacked by a Great White Shark in 1964. Back on board, his companions put a tourniquet on him (ABOVE). He continues to dive, despite having had the flipper which he had fitted to his stump ripped off by another shark. *(Photos D.R.)*

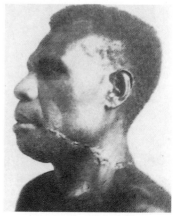

Iona Asaï is another miraculously saved. He survived the bite of a shark that tried to rip his head off, escaping with his life only because he reacted quickly to crush the shark's eyes with his two free hands. *(D.R.)*

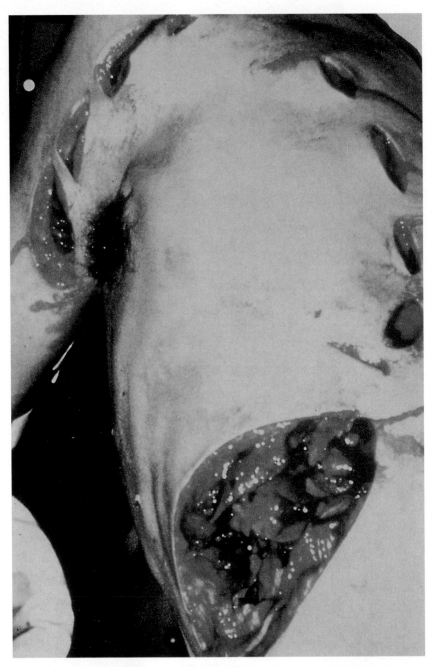

Rodney Fox is an Australian diver famous for having miraculously survived an attack. He was the victim of an "investigative bite", in other words no flesh was removed. The sum total of his injuries was nevertheless spectacular: the stomach, the left lung and the ribs were exposed, and the ribcage, the left humerus and the pneumothorax fractured. A four hour operation and 462 stitches, however, saved his life and allowed him to carry on diving. *(D.R.)*

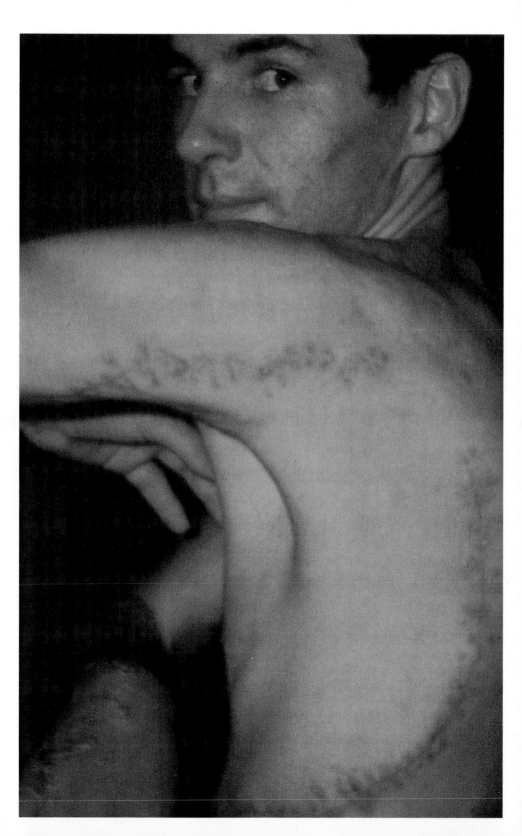

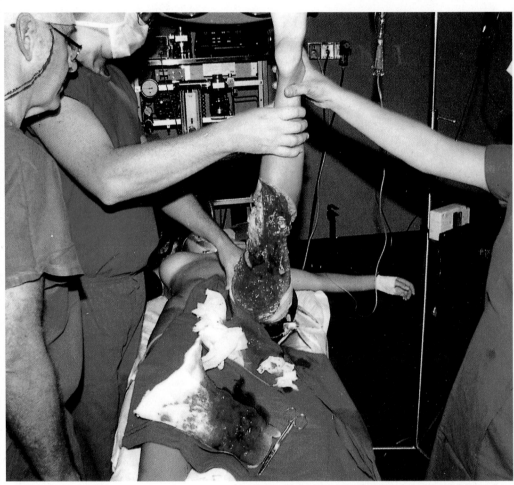

On 13 February 1988, at Mtunzini (Zululand, South Africa), 15 year-old B. had been surfing for over two hours when she decided to take a rest on her board. Suddenly she was thrown into the water and then bitten three times on the left leg. One of her companions helped her back on to her board and got her back to the beach. Having taken a first-aid course, and unable to do any tourniquet on the femoral vessels which were bleeding profusely, he decided to place a finger in the vessel to arrest the flow. This swift reaction, unique in the records, saved B.'s life. The shark responsible for the attack was a Bull Shark (*Carcharhinus leucas*), which was obviously starving, as it tore off flesh to feed. *(NSB/X.M.)*

Perfectly camouflaged and absolutely motionless, crocodiles can watch prey in this way for hours, awaiting the right moment to attack. They are dangerous predators found in mainland waters, swamps and river banks. This Nile Crocodile is very widespread throughout Africa and in particular in South Africa. *(D.R., Barrit/GAMMA)*

Discovered in the stomach of this Indo-Pacific Porosus Crocodile, captured in Sumatra (Indonesia), was the complete body of a man who had had his limbs torn off before being swallowed whole. *(D.R.)*

In the stomach of this 3.5 m Nile Crocodile captured in South Africa, a macabre surprise awaited the hunters ... *(Barrit/GAMMA)*

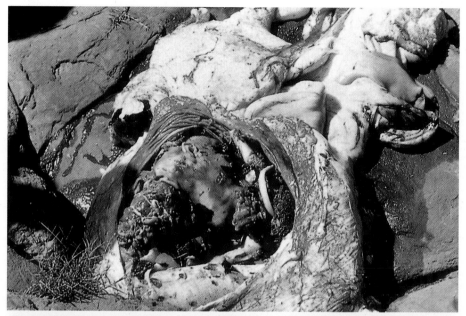

Despite its relatively small size, a crocodile such as this possesses a stomach sufficiently vast and dilatable to hold the whole of an adult body. (TOP) Human remains appear in the opening of the stomach. Unlike sharks, crocodiles only swallow human bodies they have already drowned. They then dismember them by shaking them very violently above the surface. In this case, the colour of the remains, the swelling of certain teguments and the erosion of certain tissues are evidence of several days digestion. *(Barrit/GAMMA)*

The superb beaches of the province of Natal in South Africa are often the scene of spectacular encounters and, occasionally, of deadly struggles. The notices of the Natal Shark Board are a warning to incautious swimmers or surfers. *(X.M., M. Pignieres/SIPA, NSB/X.M.)*

divers under the water, but nor is it very timid either and it likes to come and see what is going on. Before attacking, it generally circles around its victim for a while, on the whole unlike the large predators such as the Tiger or the Great White, which suddenly appear from any direction and rush straight at their victim.

As we see, the shark plays a much greater part in our daily life than we might think. Even if newspaper headlines and sensational films are only an epiphenomenon with respect to a predator that does not haunt British coasts, the shark remains a unique animal, to be taken into consideration in fields of research as varied as ecology, adaptation, evolution, hydrodynamism, physiology in general, safety at sea, etc.

8

THE OTHER JAWS OF DEATH

PIRANHAS: MYTH OR REALITY?

PIRANHAS HAVE A REPUTATION as man-eaters, but, though it is true that their jaws are equipped with sharp-edged teeth like razors, we have absolutely no reliable documented evidence proving indisputably that people may have been killed and then devoured by piranhas.

One major fact which stands out is that, unlike sharks or crocodiles, a piranha cannot kill a fully conscious man on its own. The diameter of its jaws is too small to inflict immediately fatal wounds, and it is hard to imagine a man allowing himself to be slowly nibbled to death. I am excluding, of course, any unfortunate injury to the carotid arteries, for, even in Britain, one or two deaths every year are attributable to such accidents caused by small domestic animals. The maximum size of piranhas is no more than 60 centimetres, their jaw is much less developed than that of the shark, and there are no accounts of a man being killed by a shark of less than 1.2 metres. I see in this additional proof of the impossibility of a fatal attack by a solitary piranha. A fatal attack would therefore have to be perpetrated by a large number of these fish, and in circumstances that could be compared with the feeding frenzy of sharks.

There is no doubt that piranhas in a group are capable of working away at a corpse to the point where only the skeleton is left, but this does not prove that they are capable of killing. Moreover, it is not certain that their relentlessness is the same when a prey fights back, nor any proof that their appetite is so frenzied that nothing can put them to flight.

In 1985 and 1986, Sazima and Guimaràes, two Brazilian scientists, studied three cases of necrophagy which might explain why piranhas have a reputation as "man-eaters". These three cases were observed in the Mato Grosso on three different rivers.

In the first case, a young woman of 25 drowned when she fell from a boat in mid river. Her body was recovered four days later, the detectives having spotted a bubbling in the water, caused by the gathering of piranhas. All that remained of her body was a perfectly cleaned skeleton, apart from a little flesh remaining on the left thigh.

In the second case, a 50-year-old man drowned while crossing the river on horseback. His body was found a few hours later minus the right ear and the right cheek, part of the tongue and the whole right side of the neck. This one-sidedness of the wounds was explained by the body having been resting on its left side. The rest of the body was protected by thick leather clothing as worn by the *vaqueros* of that region.

In the third case, a 70-year-old man fell in the water after having suffered a coronary thrombosis. His body was recovered twenty hours after his death with flesh remaining only on the trunk. The clothes had been torn off, except on part of the trunk. The head, the neck and the four limbs were nothing but bones, and the phalanges had been removed.

The three bodies revealed bite marks characteristic of at least two kinds of piranha. These fish belong to the Characiformes order and are found only in all the great rivers of South America. They feed on fruits, leaves and fish flesh. Many are therefore omnivorous, and those which are on the whole carnivorous content themselves with fish flesh, to the exclusion of mammals. As the above examples prove, they do on occasion eat mammal or bird flesh as carrion. This is the case, in particular, with *Pygocentrus nattereri*, which can penetrate the relatively thick skin of mammals. For the two Brazilian scientists, it is from this necrophagy that *P. nattereri* gets its reputation. If the three bodies had been recovered without the reasons for their deaths being fully known, it is certain that these incidents would have sustained the terrifying rumours of man-eating piranhas. It is significant that these two Brazilians have no evidence relating to people having been wounded, killed or devoured by piranhas. They do not, however, totally rule out the possibility that a group of *P. nattereri* can kill and eat a human being in certain circumstances,.

Such eventualities have been mentioned by authorities such as Schultz in 1964, Markl in 1972 and Goulding in 1980. The two stories related below are accepted in contemporary Brazilian history, but I pass them on to the reader without committing myself.

First is the story of a squadron of mounted police who forded a river in the 1930s. The piranhas having attacked the legs of their horses, the whole detachment was rapidly annihilated. This mythical story would, of course, only be worth giving in detail were it believable, I mention it only for information. The second story perhaps seems more plausible.

On 5th April 1961, a small launch the *Macucim* took on board about fifteen passengers on the Rio Purus, a tributary of the Amazon. The boat had been provisioning the rubber-planters in this remote semi-virgin region to the southwest of Manaus for some years. On board were some schoolgirls returning to their boarding school, some traders going to assess the price of rubber and a young couple on a honeymoon trip, José and Isaura Santiago.

Everything was fine on the small motorboat until it collided with a log of wood and sprang a leak. The damage was serious and soon proved irreversible. Nobody panicked, everyone knowing that they only needed to swim a short distance in a warm river with which they were all very familiar. The motorboat sank. The fifteen passengers ended up in the water and started to swim. Was one of them injured and losing a little blood? The water suddenly started to seethe, and they guessed that it was thousands of fish wriggling near the surface. Unfortunately they were piranhas. Within a few seconds, the torrent became tinged with red and the victims of the wreck began to scream with pain. All except a sailor who had managed to climb on to a big plank of wood from where he looked on helplessly at the nightmarish scene.

In a state of shock, he reached a village and sent an urgent cable to the authorities. A hydroplane arrived the next day, and downstream of the wreck found skeletons with whitened bones, picked clean of all flesh and held together only by a few ligaments. The account of the recovery of the skeletons is plausible, and I do have photographs of such skeletons, however I did not want to include them in the sealed section of this book, their demonstrative significance being limited. On the other hand, that all the passengers could have died such an atrocious death without more than one of them managing to reach the bank seems unlikely to me.

THE KILLER WHALE

The Killer Whale, or orca or grampus, is the only marine mammal that could be perfectly capable of being a formidable enemy of man. Its size (up to 9 metres for the male), its jumping ability (over 5 metres), its speed, its very powerful jaw, its massive pointed teeth, all in the service of one of the leading animal intelligences, could make it the equal of the most dangerous sharks. However, the improbable film *Orca* goes beyond all sense of proportion in this direction. There is no evidence that a killer whale has ever killed and eaten a human being. If that were to happen, it would probably be in the Arctic or Antarctic Oceans, much frequented by these animals which move about in schools, and there are a few stories around about Eskimos who are supposed to have been killed in this way. Human remains have been found in the stomach of at least one specimen, but there was nothing to prove that the man had been killed by the whale. This animal feeds on a variety of prey: other marine mammals including isolated whales, fish and birds, but we have no evidence to justify adding human beings to the list, especially as the Killer Whale is an "intelligent" animal whose powers of discrimination are far superior to those of the shark.

CROCODILES: FEARSOME PREDATORS

Since time immemorial all crocodiles have been considered potential predators of man. The Chinese mentioned this terrifying animal, while Marco Polo was the first in Europe to speak of these "great legged serpents capable of swallowing a man whole".

Nevertheless, although all crocodiles are capable of defending themselves aggressively, like any animal that feels threatened, only two species exist which could justifiably be considered man-eaters: the Nile Crocodile (*Crocodylus niloticus*) and the Saltwater Crocodile of the Indo-Pacific (*Crocodylus porosus*). Capable of reaching 7 metres in length and 1,000 kilos in weight, these crocodiles are responsible for numerous tragedies, usually unreported by the media, which take place in the vast African swamps, in isolated lagoons on far-off islands of the Pacific, or in the immense Australian bush. These anonymous victims are never officially recorded, but I believe their number can be estimated at several hundred per year, by extrapolating from the figures and evidence that were given to me in South Africa by certain trustworthy wardens. All the Australians know, for their part, that many Aborigines are devoured each year by the redoutable *porosus* throughout the north of the country.

Unlike what happens in general with sharks, attacks by crocodiles are more often than not isolated. So what took place in 1945 in Southeast Asia remains exceptional. On 19th February 1945, a thousand or so Japanese soldiers were trying to escape from the island of Ramree where they had been surrounded by British forces. Thirty kilometres of mangrove swamp separated them from the Burmese coast. An Englishman who was sitting in a barge grounded on the bank of a channel running through the swamp, witnessed a terrible scene: "That night was the most ghastly that the crewmen of the landing barges had ever experienced. From the pitch-black swamp came the sound of shots, punctuated by the screams of the wounded men crushed by the jaws of these enormous beasts and the muffled sounds of the crocodiles harassing them, all producing a hellish cacophony the like of which has rarely been heard on earth. Of the thousand Japanese who entered that swamp, only about twenty survivors were found." Such carnage is comparable with that of which sharks are capable at scenes of shipwrecks in tropical seas, but it is unique in modern history.

In Indonesia, Malaysia, New Guinea and northern Australia. entire communities live in fear of the Saltwater Crocodile. Millions of villagers are obliged to share this crocodile's territory and pay a very heavy price for this forced cohabitation. In the 1960s, a New Guinea missionary reported 62 villagers killed or injured by a single crocodile. On a little island off Mindanao, in the Philippines, nine fishermen have been killed in the last few years by one and the same Saltwater Crocodile.

This monster can easily reach 7 metres and weigh over 1,000 kilos, and is still very agile and fast. Its habitat is very cosmopolitan, from open sea to inland lakes by way of estuaries and large rivers. It is capable of migrating over thousands of kilometres, going to colonise the Cocos Islands, Fiji and the New Hebrides. Often these reptiles are found covered with barnacles like old boat hulls. They lay 60 to 80 eggs every year on mounds. Though the young feed on crustaceans, insects and snakes, the adult eats any living animal within reach, fish, birds or mammals. Man is just another prey no different, for example, from the pig, and he can easily be seized by any part of the body and dragged into the water. Water remains the crocodile's favourite territory, particularly when it is preparing to attack. It is capable of keeping watch, motionless as a tree trunk, for some hours, then charging at incredible speed, leaving the water like a missile. A very great number of fishermen are grabbed like this when they are two or three metres from the waterside, having been unable to see the killer approach beneath the surface. When the crocodile launches itself, it is too late. I have seen monsters like this lie motionless for hours, and then launch themselves sideways at an incredible speed to grab a piece of meat which I threw quite far away from them. No dog is capable of such speed and such precision.

A crocodile kills its prey by crushing it in its jaws, which have a very powerful shutting action, or by drowning it. Once its victim is killed, it sets about the task of swallowing it. If the body is too big for the operation to be accomplished in a single go, the crocodile dismembers it by means of violent head movements. The body is held above the water, and is literally smashed to pieces in this manner (see photos).

Though it is true that these animals may feed on carrion, they always prefer fresh flesh on which they gorge themselves without limit, and, although a Saltwater Crocodile can go up to two years without feeding, it is better to take care to avoid passing within reach of it. Again because they prefer fresh meat to carrion, it is now not certain whether the theory that they store their catches under tree roots while waiting for them to be softened up by the decay process is correct. They are powerful enough to dismember and decapitate a large mammal immediately after killing it.

Statistics from Australia show that attacks take place during the hottest months, corresponding with an increased activity. It is in fact at this time that the crocodiles feed and mate. Out of 27 fatal attacks which it has been possible to study, 16 victims were swallowed entirely, while eight disappeared and three others were found intact. In 50% of cases, crocodiles longer than 4 metres (and therefore males) were involved.

A crocodile which has captured and eaten a human being will very often have a tendency to return to the same spot to do it again. This observation explains certain serial attacks and justifies the hunting of known aggressors. In India, where the Saltwater Crocodiles, and even the gavials, were used to eating human remains as a result of religious customs, attacks on living people were very frequent. They are less so now owing to a major decline in the population of crocodiles in that region.

The crocodile's metabolism produces an excellent yield, enabling it to fast for very long periods. Its stomach, particularly acid, allows it to digest the whole of its prey, including the bones, and the energy of the metabolites is stored in the form of fat in the tail and along the back. It is thanks to these reserves that an adult weighing one tonne can fast for two years.

In Zaïre, on 12th July 1987, the ferry Maria was moving down the Luapula river with 500 passengers on board when the pilot fell asleep and ran his old boat aground on a sandbank. The Maria heeled over under the pressure of the current, throwing the sleeping men on top of each other among the crates of beer, sacks of flour and livestock. It finally capsized and then sank within a few minutes. It was 3 o'clock in the morning. Of the passengers, many of whom were drunk, 400 were to disappear. A terrible death awaited a large number of them via the crocodile. This region in fact is home to Nile

Crocodiles which easily reach 6.5 metres and 1,000 kilos. Less impressive than their coastal cousins, they are none the less formidable, moving very swiftly underwater in groups of several tens. Many passengers must have been grabbed, drowned and devoured by a monster which they could not even see. Countless half-dismembered bodies were discovered in the course of the following weeks.

The Nile Crocodile is found throughout Africa south and east of the Sahara. It is unarguably the most widespread predator in the whole continent, as abundant on its own as all the others put together (lion, panther, hipppotamus, rhinoceros, elephant, buffalo). The number of attacks for which it is responsible is in parallel considerable. Its mode of attacking is particularly difficult to detect owing to its perfect mimesis, a unique posture when hiding, and an approach as discreet as it is effective. With only its nostrils and its eyes above the surface, it watches for hours until spotting its prey. It then dives unobtrusively and swims rapidly towards its target without losing direction. It is only when it gets a few metres from its unsuspecting victim that it emerges from the surface in a huge leap that can propel it over several times its own length, thanks to the power of its tail and the thrust of its hindfeet on the river bed or the bank. This is how the unfortunate fisherman, the photos of whose body appear in this book, was caught out; sitting more than 1.5 metres above the surface, he could not have imagined that a crocodile could come and surprise him at that height.

Silence, speed and surprise are the constant features of all attacks. Out of a sample of 43 attacks in Zululand and in Mozambique, 30 took place between November and April, the summer period of breeding and greater activity for cold-blooded animals such as crocodiles. Several attacks have been made on individuals selected from among large noisy groups, confirming that noise and being in a crowd are but illusory means of protection. Like sharks, the crocodile is attracted from a long way off by the noise from a shoal of leaping fish or from an animal splashing about in the water.

In 1987, Gustav Giltzman was doing some underwater fishing in Lake Kariba. He came back up with some fish hooked on to his belt and surfaced among some half-submerged tree trunks. "As I was surfacing, I felt a big crocodile rub against my side. I then stuck the point of my harpoon gun into the animal as hard as I could and turned to climb on to a dead tree beside me. As I grasped the tree, the 4 metre crocodile attacked, biting my leg and shaking me like a leaf. I clung desperately to the tree and felt the animal tear off a mouthful of my thigh. I think my diving suit saved me from being bitten more badly. Before it could catch me a second time, I climbed on to the tree, the croc only just missing me as it leaped more than a metre out of the water. My nephew saved me by shouting to me not to climb any higher, as the branches were rotten." Gustav was extricated from his awkward position by the crew of an outboard who took him to Kariba where he was operated on. That same year, more than 500 crocodile attacks were recorded in Zimbabwe alone, and the Zambeze river has become one of the most dangerous on the planet, as sharks come up it and the Nile Crocodiles are invading it more and more.

In Florida, the American alligator makes about fifteen attacks on man every year, and complaints made to the game-fishing board are increasingly numerous (several thousands annually). A young girl of 16 was killed in August 1973 while swimming with her father in the Sarasota district; an undersea diver was killed in 1987 at a seaside resort in the north; and a young child was killed in 1988, a stone's throw from a residential area. Even though it feeds mainly on fish, amphibians, birds and small mammals, it remains an opportunist and on occasion attacks dogs, calves, pigs, goats, children and small adults. Unlike the other species, the female alligator defends its nest by giving many and varied warnings: hissing, opening the mouth, raising the tail out of the water, inflating the body. All this behaviour allows one to take shelter before being attacked.

Before 1969, the alligator was hunted prolifically and so avoided man. Since it has been given protection and children feed it near villages, it is no longer shy and the authorities are wondering whether this factor might not encourage attacks.

Those crocodiles said to be man-eaters and the dangerous sharks often frequent the same geographical areas, South Africa, Florida, Australia. So, quite understandably, some disappearances are sometimes blamed on one or the other, with nobody being able to say for sure which one was responsible. On 21st October 1978, Mr Kuppasamy left to go fishing 245 kilometres north of Durban, on a river named the Saint Lucia, very near its estuary. His fishing tackle was found on a bank, with footprints disappearing into the water, but there was no sign of the fisherman at all. Five days later, his body was found upriver, mostly eaten. Although it is probable that a number of predators subsequently attacked the dead body, the investigators were unable to give a verdict on the killer's identity, Tiger Shark or Nile Crocodile.

ELEMENTARY PRECAUTIONS

The risk of meeting a crocodile is greater than the risk of a tête-à-tête with a shark, for we often share terrestrial habitats with the former.

- Find out from the local inhabitants about the habits of the crocodiles and where they lie up. If you go fishing or hunting, you have every chance of encroaching on their territory. Defence of the territory is the essential attribute of the male against any intruder, even a man, or a boat. The female defends only her nest, and she can bite if given the chance, even though no known fatal attack can be attributed to her. A canoe resembles a rival in their territory, and the male will often attack this type of boat.

- Do not trail an arm or a leg in or even above the water. On a boat, do not sit on the gunnel.

- If you are fishing, do not throw the entrails of your catches into the water and install yourself at a minimum distance of 3 metres from the edge.

- A newborn young is never far from its parents, so do not approach it.

- Near the water at night, always have a torch with you or a cap lamp on your forehead. This is indispensable for hunting crocodiles, whose eyes shine a very characteristic bright orange.

- Just because you don't see the orange eyes of crocodiles does not mean that they are not there, but simply that they are hiding.

- Crocodiles are attracted by noise. It is therefore a poor deterrent. So firing a gun into the water when crossing a ford, does not guarantee the required result, although you might always put a shot in your assailant's eye.

- Be wary of harmless-looking stagnant pools, they can conceal large predators, even in 30 centimetres of water. Having had the chance to hunt crocodiles on horseback in Colombia in the 1970s, I was surprised at the size of some animals in relation to the pool where I shot them. I did not notice it until, hit in the head, they generally made one final leap before coming to rest on their back out of the water. At night, my horse would warn me if a crocodile was nearby on land.

Having also hunted for food, small crocodiles (Cuvier's Smooth-fronted Caiman) in the interior of Guyana, upstream of Maripasoula,

I have been able to verify certain simple rules. These crocodiles were no longer than 1.5 metres and so were not really dangerous, provided they were killed before being put in the bottom of the dugout canoe. To kill them means putting them out of their misery with a machete, not only in the nape area by cutting half through the neck, but also in the tail area level with the hindlegs. I have seen crocs, even when slashed like this, lash out violently with the tail an hour after being captured, and viciously bite anything that came into their clutches. A 1 metre specimen made score marks with its teeth on a machete which I put into its mouth, the same machete that had half decapitated it and cut through its tail. If I add that I had shot it straight in the head with a bow and arrow before hoisting it aboard, you have an idea of the incredible vitality of these primitive animals, which still have reflexes for biting even when the spinal cord is cut at the back of the neck. A dead crocodile is a crocodile cut up into pieces.

If you are grabbed by a crocodile and you have time to think, you should know that your only chance (if you are unarmed) is not to allow yourself to be dragged into the water. An adult croc can be fifteen times the weight of a man, and the latter has no chance when out of his element. With an attacker up to 2.5 metres in length, a man has a chance of escaping or of surviving his injuries (2.5 metres corresponds to about 100 kilos in weight). It must be remembered that, unlike sharks, the crocodile does not kill through wounding and haemorrhage, but more often than not by drowning, and much more rarely by crushing vital organs. A French soldier was once very lucky and at the same time very quick to react when he was pulled by the arm into the water. His arm was completely enclosed in the mouth of a Nile Crocodile, which began to sink in the water. The soldier then caught hold of the enormous valve-like glottis that seals off the rear of the animal's mouth and prevents it drowning by stopping water coming in. As soon as he managed to catch hold of the organ with his free hand and pull it, the monster immediately let go of him. This anecdote should not be seen as a general rule, but only as a reminder that no situation is ever completely hopeless, no matter how desperate it may appear.

BARRACUDAS: UNJUSTLY ACCUSED

If there is one animal which is unjustly accused of the worst acts of brutality towards man, it is the barracuda. Hans Haas had no hesitation in stating dogmatically that there was no doubt that the barracuda attacked, and on occasion killed, human beings. In Brazil, the barracuda is even more feared than the shark, and in the Caribbean it is considered dangerous and particularly fierce when caught in a net. It is often called the "tiger fish" or the "long-nosed sea pick", in reference to its long snout extending from a long-line body. In June 1922, Dorothy MacLathie was bitten to death when bathing at St Petersburg on the west coast of Florida and barracudas were blamed. On 4th August 1947, a student was killed by a fish at St Augustine, again in Florida: once more a barracuda was accused. In both cases, the morphology of the injuries clearly indicated an attack by a shark, not a barracuda. The first inflicts a wound that is circular and highly characteristic, while that of the second is straight and punctuated. Any "expert" with the slightest experience could have made the diagnosis, but this "miscarriage of justice" was to suit the local authorities for obvious reasons of tourism. The word "shark" already had, at that time, far more sensational repercussions than the word "barracuda". It is significant, incidentally, that nobody ever dreams of blaming this animal in Australia, although it is just as abundant there.

In Florida, again, and not so long ago, they even went so far as to assert that if there were no barracudas in the local waters there were no shark attacks. The reason given was that sharks only attacked human beings already wounded by barracudas. It is just not credible.

Let us be serious: even though the barracuda does not have a very pleasant mouth, with two rows of pointed teeth in the upper jaw and another in the lower jaw, it never attacks if it is left alone. I have seen hundreds of them in the Caribbean and in the Pacific, and they ignored me in the same way as I ignored them. At the very most they came to look at me from a few metres distance, approaching slowly, from the front, with no threatening movements, out of a simple curiosity which I would be tempted to qualify as "kindly" were this not anthropomorphic. These majestic fish often go around in shoals of four or five (I have counted up to nine of them), in "forward-inclined stepped formation" in the vertical plane, a few metres beneath the surface, and are often stationary, being interested only in those fish which they are capable of swallowing. Even for the biggest among them (around 1.8 metres), neither the diameter and the shape of their jaw nor the shape of their teeth would allow them to tear off segments of limbs and to cut through bones the way sharks do. People who denounce them are no doubt forgetting that the biggest barracuda caught was in 1932 in the Bahamas: it weighed 50 kilos and measured barely 1.8 metres.

Barracudas are no more aggressive than moray eels, which never attack first if they are left undisturbed inside their dens. Man is too inclined to forget that, if there is a single point in common between the animal world and himself, it is the instinct of self-preservation.

THE JAWS OF DEATH

This genetically inscribed instinct will make a barracuda pierced by a harpoon dart try to bite the hand that comes near it, a cornered moray eel seize the arm that threatens it, and a peaceful shark dismember the irresponsible person who attempts to ride it.

9

DIRECTORY

THE 343 OR SO SPECIES OF SHARK currently recognised are divided into 8 orders and 31 families. I do not think it is relevant in this book which is intended above all to be practical, to give a strict description of the method of classification, or to draw up an exhaustive list of all the shark species in the world. For those who wish to get hold of such a catalogue, there is a real "bible" in this field, drawn up on behalf of the FAO (Food and Agriculture Organisation) by Professor Compagno, the world's leading shark expert (see *Bibliography*).

The directory which follows does, on the other hand, include all the 31 species which attack, have attacked or are liable to attack man. I have added to these a few species which do not present any risk, either because they are of interest for the study of sharks in general or because they show the immense polymorphism of sharks. Each species is depicted by several diagrams:

- general profile;
- underside of the head; upper and lower dentition;
- cutaneous denticles (if characteristic);
- geographical distribution.

The text is restricted to the most interesting features. The name in capitals is the English name, this being followed by commonly used alternative names where appropriate, and beneath this the scientific name in italics. The same shark can be called by different or even conflicting vernacular names depending on the region of the world you are in. Different sharks may also be referred to by an identical name in two different countries: thus, in Australia, "whaler shark" is often used to denote all the Carcharhiniformes, in other words most of the sharks liable to attack man do so for the same reason as they attack whales.

There are 8 major groups or orders, identifiable by simple morphological characteristics, within the grasp of any observer as the following directory illustrates. If, one day, you come into contact with a big shark which does not have any anal fin, but whose body is not flattened and whose snout is short, you will be absolutely correct to assume that it is a shark of the order Squaliformes. This order comprises 3 families and 71 species, which may not be of any interest to the non-specialist, but if you are north of 45°N you will be able to call to mind the Greenland Shark, the only one within these three families that is potentially dangerous.

The truly dangerous sharks as a whole belong within two orders only: the Carcharhiniformes and the Lamniformes. The other orders include a fewer number of potentially dangerous species. As already mentioned, there is one species in the Squaliformes (the Greenland Shark). There are two species in the Orectolobiformes (Spotted Wobbegong and Nurse Shark). And there is one species in the Hexanchiformes (Broadnose Seven-gill Shark). The three remaining orders (Squatiniformes, Pristiophoriformes, Heterodontiformes) do not contain any dangerous species. The *very* dangerous sharks are four in number:

- the Great White Shark;
- the Tiger Shark;
- the Bull Shark;
- the Oceanic White-tip.

The first is a lamniform and the other three belong to the Carcharhiniformes.

All the sharks that have attacked, injured or killed man are indicated by one, two or three death's-head emblems according to their level of danger. I have not taken into consideration sharks which have killed by accident because they were clumsily or imprudently handled when outside their natural environment. Thus, the Thresher Shark is not dangerous, even though it managed one day to decapitate an unfortunate sailor on the deck of a ship; any more than is the Whale Shark which crushed another sailor when its dead body rolled along the deck of a whaleboat. I shall mention only for the record, the case of the Scottish fisherman who, in 1960, got his arm bitten by a harmless shark which he had just pulled up on to the deck of his boat. This incident, like those which happen in their dozens every day with dozens of different fish, nevertheless appears in the annals as the most northerly shark attack ever recorded. Even if we talk of a provoked attack (a bit of a euphemism), it is not a question of aggression but of the most basic instinct of self-preservation.

The copyright in the figures contained in this directory belongs to the Food and Agriculture Organisation of the United Nations (FAO), which has kindly given the author permission to reproduce them. These figures are taken from volume 4, parts I and II, of the *Fisheries Synopsis* no. 125, entitled *Sharks of the World*.

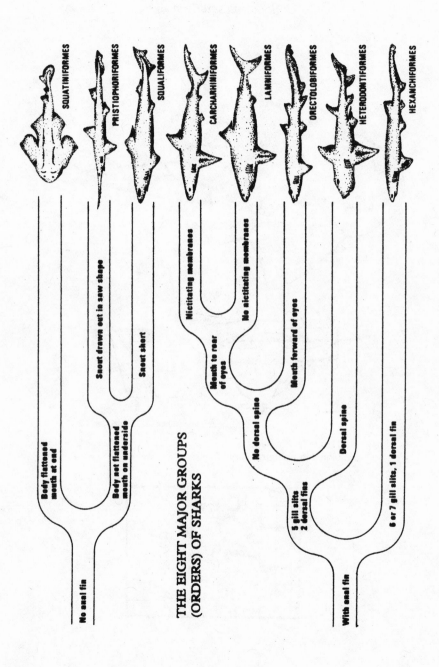

THE EIGHT MAJOR GROUPS
(ORDERS) OF SHARKS

SQUATINIFORMES
PRISTIOPHORIFORMES
SQUALIFORMES
CARCHARHINIFORMES
LAMNIFORMES
ORECTOLOBIFORMES
HETERODONTIFORMES
HEXANCHIFORMES

Body flattened
mouth at end

Body not flattened
mouth on underside

Snout drawn out in saw shape

Snout short

Nictitating membranes

No nictitating membranes

Mouth to rear
of eyes

Mouth forward of eyes

No dorsal spine

Dorsal spine

No anal fin

5 gill slits
2 dorsal fins

With anal fin

6 or 7 gill slits, 1 dorsal fin

ANGEL SHARK (Monkfish)

Squatina squatina

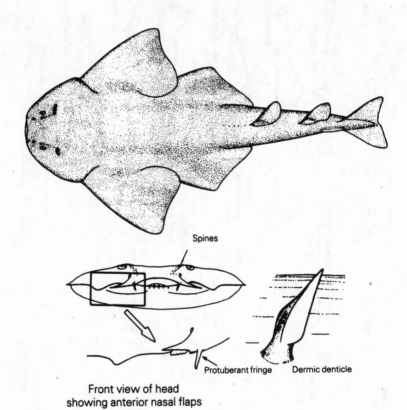

Spines

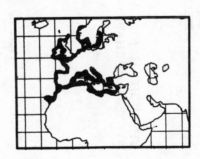

Protuberant fringe   Dermic denticle

Front view of head
showing anterior nasal flaps

ORDER: Squatiniformes.

FEATURES: Small spines from head to tail.

HABITAT: A shark from temperate sea beds of Europe, from the coast to depths of 150 metres. Prefers muddy or sandy bottoms, where it lies with only its eyes showing above the mud. Nocturnal activity. Sleeps on sea bed during day. Penetrates farther north in summer.

REPRODUCTION: Ovoviviparous.

FOOD: Bony fish, skates/rays and other flatfish, crustaceans and molluscs.

MAXIMUM SIZE: 1.83 to 2.44 metres.

BAHAMAS SAWSHARK (American Sawshark)

Pristiophorus schroederi

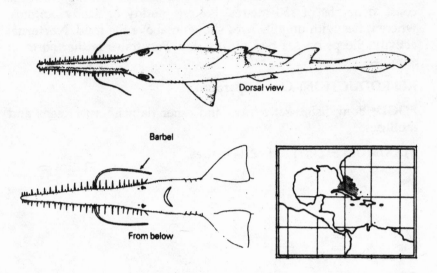

Dorsal view

Barbel

From below

ORDER: Pristiophoriformes.

FEATURES: Saw: one-third of total length.

HABITAT: Ocean bottom around 600 to 900 metres.

MAXIMUM SIZE: 80 centimetres at least.

GENERAL: Another example of polymorphism. The sawshark of Lake Nicaragua is a close cousin of this one.

COOKIE-CUTTER SHARK

Isistius brasiliensis

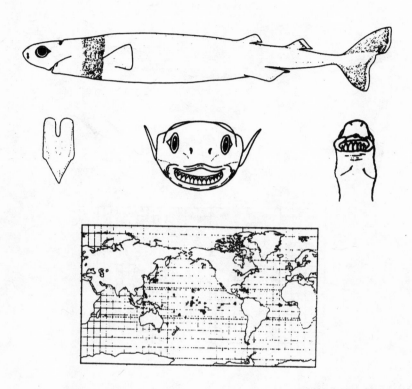

ORDER: Squaliformes.

FEATURES: Jaws and teeth very powerful. May bite fishermen once captured, for becomes active and swift. Caught at night, its body is covered with luminescent organs. This shark swallows the teeth it loses, perhaps to maintain its calcium level.

HABITAT: Pelagic oceanic. Vertical migrant: at night rises towards surface and by day descends more than 2,000 or 3,000 metres (contains a lot of oil in a very large liver).

MAXIMUM SIZE: 50 centimetres.

GENERAL: Leaves tooth marks in the skin of the big marine mammals and in the plasticised or rubber covering of submarines.

GREENLAND SHARK   (Sleeper Shark)
Somniosus microcephalus

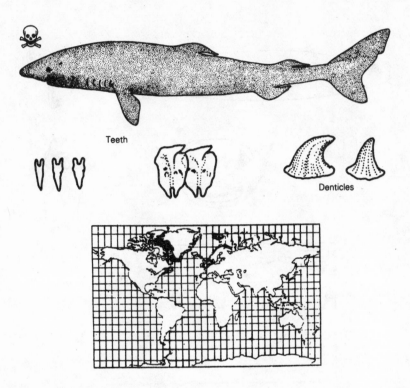

Teeth

Denticles

ORDER: Squaliformes.

FEATURES: Heavy cylindrical body. Teeth 52/48.

HABITAT: Largest fish of arctic Atlantic and of Antarctic, seen at increasing depth as temperature rises. Its ideal habitat temperature is from 6°C to 12°C.

FOOD: Fish but also marine mammals (seals in particular) and all kinds of carrion. Seems impervious to gunshots and cutting instruments when gorging itself on food. Parts of a horse and an entire reindeer have been found in big Greenland Sharks. Luminescent copepods live as parasites on each of the shark's eyes and may act as lures for potential prey. This shark is said to have overturned Eskimos in their kayaks and eaten them, but no reliable documentation has ever confirmed this.

MAXIMUM SIZE: 7.3 metres.

GENERAL: Proverbially sluggish; puts up hardly any resistance to capture. Easily fished in the Arctic through holes in the ice. The Eskimos use its skin to make boots and its teeth to make knives for cutting hair. The fresh flesh is toxic except if it is washed carefully, and is eatable dried or semi-putrefied.

PIKED DOGFISH (Spiny Dogfish, Spurdog)
Squalus acanthias

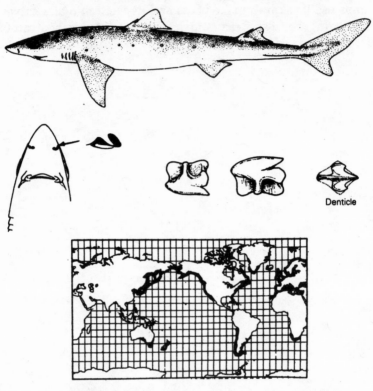

Denticle

ORDER: Squaliformes.

FEATURES: Grey above with white spots, and white below. Very widely distributed.

HABITAT: Boreal to warm temperatures. From the surface to over 900 metres depth. Seen more often near bottom. Without doubt the mostwidespread shark in the world, it is fished industrially. 35,000 tonnes of this fish were caught in 1978 in Europe. Slow and inactive, it can gather in enormous shoals. In the northwest Atlantic, lines with 700 to 1,500 hooks have been brought up with a dogfish on just about every hook. It does not tolerate non-saline water and moves away from coasts at times of heavy rain. Its ideal water temperature is from 7°C to 15°C, explaining its seasonal migrations downwards or northwards (a tagged individual was caught again seven years later 6,500 kilometres away).

REPRODUCTION: Ovoviviparous. Gestation lasts 18 to 24 months. Between one and 20 young born according to size of the mother. Maturation: 14 years for the male, 23 for the female, with an average maximum age of 30 years and some estimates of up to 100 years.

FOOD: Voracious predator, eating all kinds of fish and invertebrates. Not dangerous but, for fishermen, its pointed teeth and its spines can cause serious wounds and allergic reactions. It is the prey of numerous big sharks, seals and bony fish.

MAXIMUM SIZE: Not more than 1.6 metres.

SAILFIN ROUGHSHARK (Humantin)

Oxynotus paradoxus

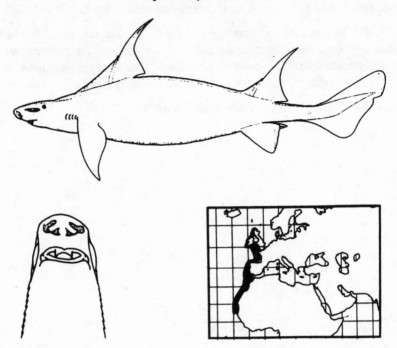

ORDER: Squaliformes.

FEATURES: Uniform dark grey or dark chestnut. Dorsal fins shaped like sails with spines.

HABITAT: Found on Atlantic coasts of Europe (France in particular).

GENERAL: Ovoviviparous.

MAXIMUM SIZE: 1.18 metres.

GENERAL: A good example of the polymorphism of sharks.

SILVERTIP SHARK

Carcharhinus albimarginatus

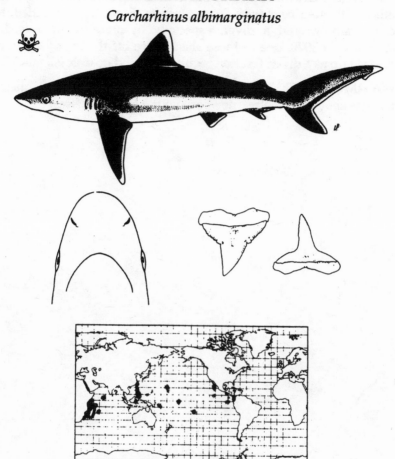

ORDER: Carcharhiniformes.

FEATURES: Dark grey above, bronzy-white below. All the tips of the fins and the rear margins are white. Faint white band on the side.

HABITAT: Pelagic, with liking for banks, reefs or islands in open ocean. Lives mainly between the surface and 600 to 800 metres depth. Often follows boats on the high seas. The young grow up in shallow waters near coasts.

REPRODUCTION: Viviparous. The young (1 to 11 per litter) are born in summer after gestation period of 1 year.

FOOD: Open-water and bottom-living fish (flying fish, tunny, bonito, sole, octopus, skate). Equal in size, but more aggressive than

Galapagos Shark and Black-tip Shark in taking bait. Adults fight each other and often bear scars from conflicts. Few attacks verified, but can fatally wound a diver, especially in presence of a feeding stimulus. In 1970, one of these sharks tore off the leg of a dummy dressed up as a diver. Caution required around oceanic islands.

MAXIMUM SIZE: 3 metres. Adult males around 1.6 metres, females about 1.70 metres, newborn young 65 centimetres.

STARRY SMOOTHHOUND (Sweet William)
Mustelus asterias

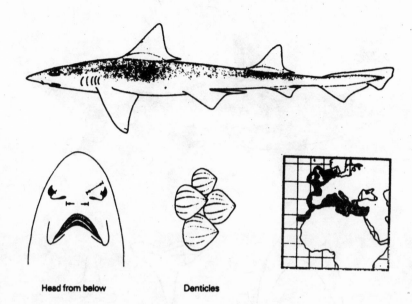

Head from below   Denticles

ORDER: Carcharhiniformes.

FEATURES: Spotted with white.

REPRODUCTION: Ovoviviparous. Gestation period 12 months, 7 to 15 pups per litter. Maturation rapid, in 2 to 3 years.

FOOD: Crustaceans and hermit crabs (together with shells).

MAXIMUM SIZE: 1.4 metres.

GENERAL: The flesh is consumed fresh and salted. This shark presents no danger.

GRACEFUL SHARK

Carcharhinus amblyrhynchoides

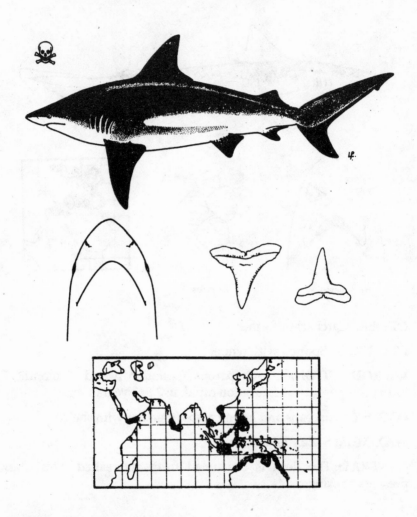

ORDER: Carcharhiniformes.

FEATURES: A grey pot-bellied shark. Tips of fins generally black. Faint white band on flanks. 15 rows of teeth.

HABITAT: Pelagic and coastal.

REPRODUCTION: Viviparous.

FOOD: Mainly fish, like its congeners. No attack recorded so far, but potentially dangerous on account of its size, its teeth, its habitat and its feeding habits. Still little known about this species.

MAXIMUM SIZE: 1.67 metres (adult female); size of newborn pup 52 to 55 cm.

GREY REEF SHARK

Carcharhinus amblyrhynchos

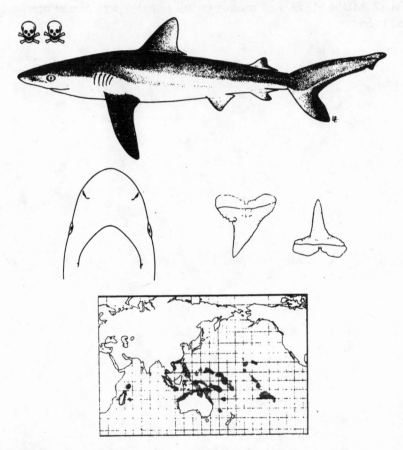

ORDER: Carcharhiniformes.

FEATURES: Grey, white below, large circular eyes, black band on outer margin of caudal fin. Teeth 14/13 in each half-jaw.

HABITAT: Pelagic and coastal. Common around atolls, coral reefs and near strong currents in lagoon. Lives between the surface and 100 metres depth. With the Black-tip and the White-tip Reef, most common of the reef sharks in the Pacific. Active, powerful swimmer, most active at night.

REPRODUCTION: Viviparous. Adult at 7 years. Maximum age at least 25 years.

FOOD: Reef fish of less 30 centimetres, crabs, crayfish, octopuses and shrimps. Inquisitive, inspects divers, practises "intimidation swimming". Attacks divers when wounded fish present, not always discriminating between the two. At least one fatal case reported. (See text for experiments in one-man and two-man submersible.)

MAXIMUM SIZE: 2.55 m. Adult usually 1.35 metres, newborn pup 45 to 60 cm.

PIGEYE SHARK
Carcharhinus amboinensis

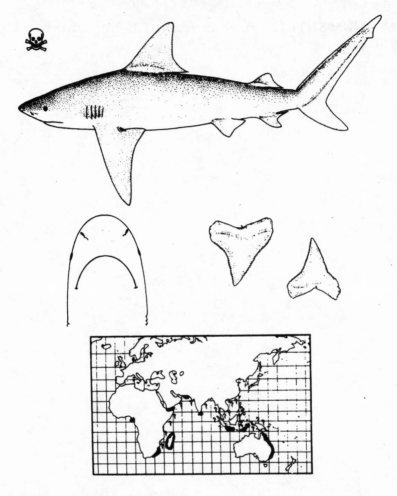

ORDER: Carcharhiniformes.

FEATURES: Grey, small eyes, pale below. Teeth 12/11.

HABITAT: Coastal species, 0-60 m. depth, along beaches and in surf zone. Off Madagascar more abundant than Bull Shark, and inversely off Mozambique. Possible mutual exclusion between the two through competition. Fished for on long lines in western Indian Ocean for human consumption of the fresh flesh.

REPRODUCTION: Viviparous.

FOOD: Mainly bottom-living fish, crustaceans and molluscs. No attack recorded, but according to Professor Compagno must be considered potentially dangerous (size, large jaws, teeth).

MAXIMUM SIZE: 2.80 m. Adult males 1.95 m, females 2.10 m, newborn 71-72 cm

COPPER SHARK (New Zealand or Bronze Whaler, Narrowtooth Shark)
Carcharhinus brachyurus

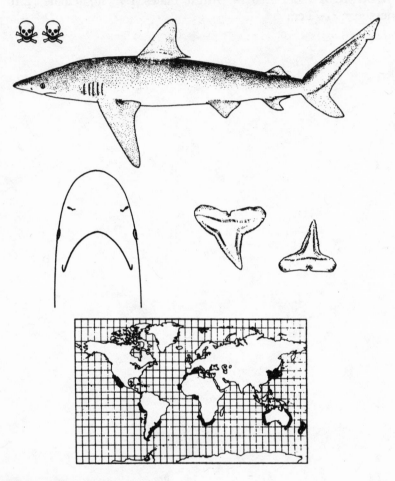

ORDER: Carcharhiniformes.

FEATURES: Bronze to olive-grey above, white below. Teeth 15-16/15.

HABITAT: Coastal, from surf to 100 m. depth. Migrates N in summer, S in winter.

REPRODUCTION: Viviparous. 13 to 20 per litter. Sexually mature at 5 years. Maximum age at least 12 years.

FOOD: Bony fish, including sardine, mullet, gurnard, hake, and sole, electric ray, sawfish, squid, razorfish. Follow the sardine shoals along the Natal coast (see photos). Dangerous, provoked and unprovoked attacks on swimmers and divers. Less so no doubt than the Tiger or the Bull Shark, owing to its feeding habits.

MAXIMUM SIZE: 2.92 metres. Maturity: 2 to 2.29 metres for males, up to 2.40 metres for females. Newborn pups 59 to 67 centimetres.

SPINNER SHARK
Carcharhinus brevipinna

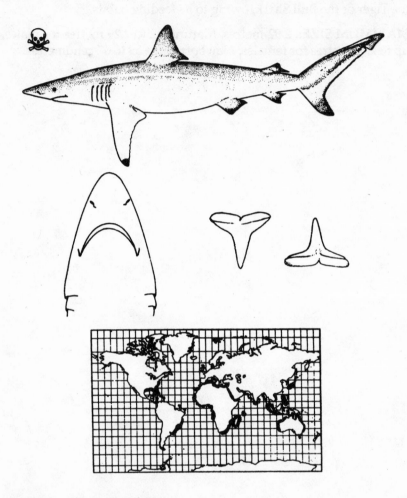

ORDER: Carcharhiniformes.

REPRODUCTION: Pale grey with a thin white line along side, unobtrusive. Tips of fins black, especially in young. Small eyes. Teeth 16/15.

HABITAT: Coastal and pelagic.

REPRODUCTION: Viviparous. 3-15 per litter. The bigger the females, the more embryos. Gestation 12-15 months. Breeds off Natal coast: females live there all year; males arrive in summer, the young prefer lower temperatures, and migrate south, towards Cape.

FOOD: Similar diet to congeners. Distinctive hunting: enters shoals of small bony fish and swims fast among them, mouth open, spinning around its big axis and snatching in all directions; ends by leaping out of the water in its momentum. Has attacked at least once, but is not very dangerous. Small, narrow teeth not suited to big prey.

MAXIMUM SIZE: 2.78 metres. Maturity: male 1.59 to 2.03 m., female 1.7 to 2 m. Newborn pups 60 to 75 centimetres.

GENERAL: Exploited for salty flesh, skin, fins, and oil in its liver (for vitamins).

SILKY SHARK

Carcharhinus falciformis

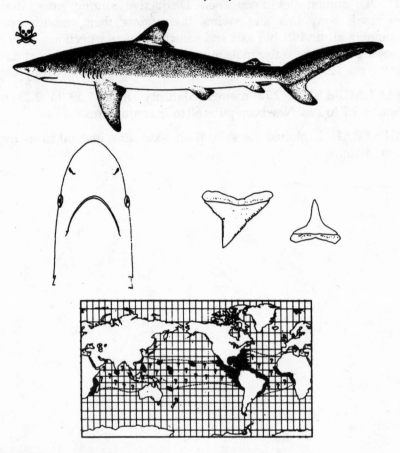

ORDER: Carcharhiniformes.

HABITAT: Mainly oceanic. Tropical, active, aggressive. Intimidation swimming given in presence of divers. One of the three most common oceanic sharks together with the Blue and the Oceanic White-tip. Known as "the fillet-eater" in the east Pacific as it often accompanies the tunny shoals. Potentially dangerous. Has attacked survivors of air disasters (as has the Oceanic White-tip).

MAXIMUM SIZE: 3.3 metres.

GALAPAGOS SHARK (Grey Reef Whaler)
Carcharhinus galapagensis

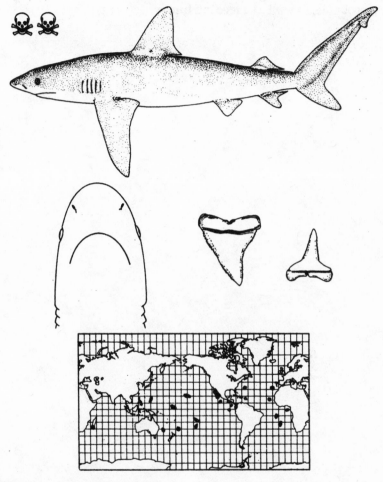

ORDER: Carcharhiniformes.

FEATURES: Teeth: 14/14. Brown-grey above, white below.

HABITAT: Not oceanic but coastal and pelagic. Lives mainly around tropical islands. Prefers clear waters, rocky beds and uneven coral.

MAXIMUM SIZE: 3.7 m. Mature males 1.7 to 2.35 m., females 2.35 m.

GENERAL: This shark may perform display swimming recalling that of the Grey Reef Shark, announcing the possibility of an attack. Aggressive, has killed a swimmer in the Virgin Islands and has attacked others, particularly in Bermuda. Diving when several of these sharks are around and a food stimulus is present makes attack very probable, as well as a feeding frenzy.

BLACK-TIP SHARK
Carcharhinus limbatus

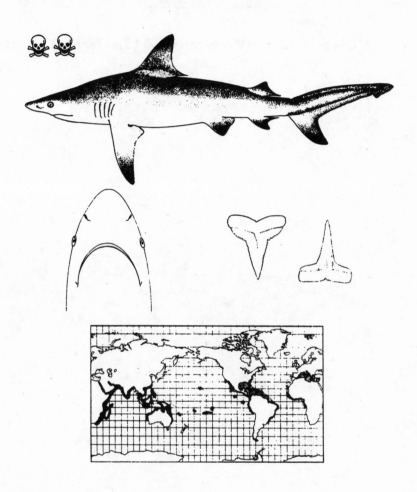

ORDER: Carcharhiniformes.

FEATURES: A grey shark with black tips to the fins in the adult and the young. Faint white band on side. A very swift shark, active, often gathering in groups at the surface. Jumps out of the water and can turn on itself up to three times before re-entering (like the Spinner Shark when feeding on shoals of small fish).

HABITAT: Not really oceanic, often near coasts, estuaries, in hypersaline mangroves, in lagoons. Does not come up rivers.

FOOD: Fish, crustaceans, cephalopods, small sharks. Subject to feeding frenzies when in a group around a definite prey. Attacks underwater fishermen carrying their catches. Less aggressive, however, than the Galapagos or the Silvertip. Its rapid speed makes it a difficult aggressor.

MAXIMUM SIZE: 2.55 m. Mature males 1.35-1.8 m., females 1.2-1.9 m.

BULL SHARK
(Zambezi Shark)
Carcharhinus leucas

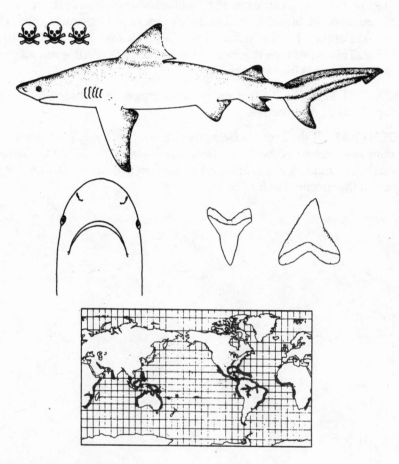

ORDER: Carcharhiniformes.

FEATURES: Teeth 13/12. Grey with tips of fins slightly darker.

HABITAT: Continental coasts of all tropical and subtropical seas; the only shark to come a very long way up into rivers and lakes: the Amazon, Mississippi, River Gambia, Zambezi, Limpopo, Tigris, Chatt-al-Arab, Ganges, Lake Nicaragua, Lake Yzabel (Guatemala), Panama Canal, Lake Jamoer (New Guinea) and Lake Macquarie (Australia). Often found in muddy waters, but also in hyposaline or hypersaline waters. Not found in any lake not connected to the sea (for reasons of breeding). Ascends the Amazon as far as Peru (3,700 kilometres). Tolerates the hypersalinity of St Lucia lake (53% instead of 35%) in South Africa. Has survived for 15 years in captivity.

FOOD: Very varied like that of the Tiger Shark, from turtles through all saltwater and freshwater fish to other sharks. Even eats young of its own species, birds, dolphins, antelopes, rats, dogs and sloths. The only difference in diet from Tiger Shark is that it is less fond of refuse. This is perhaps the most dangerous shark of all, even more so than the Great White and the Tiger.

MAXIMUM SIZE: 3.4 metres. Females bigger than males. Newborn pups 56 to 81 centimetres.

GENERAL: Fished for its flesh, its skin and its liver. This shark has often been confused with the Java Shark and especially the Ganges Shark. The many victims credited to the latter could be attributable in part to the formidable Bull Shark.

CARIBBEAN REEF SHARK
Carcharhinus perezi

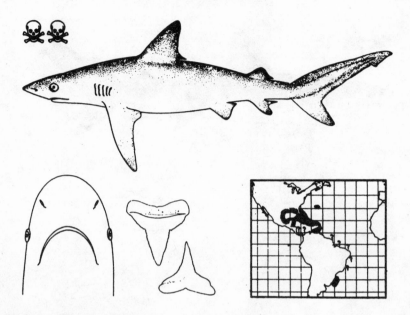

ORDER: Carcharhiniformes.

FEATURES: Plain grey or grey-brown appearance, white below.

HABITAT: The commonest coral-reef shark in the Caribbean. May stay on the bottom without moving (with its pharynx and gills pumping water). Little known despite its abundance. Has attacked two divers, and no doubt others in the past.

MAXIMUM SIZE: 2.95 metres.

OCEANIC WHITE-TIP SHARK

Carcharhinus longimanus

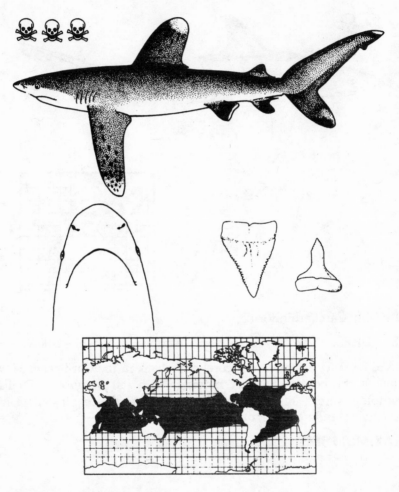

ORDER: Carcharhiniformes.

FEATURES: Easily recognised. Very large pectoral fins and tall first dorsal fin, all with white tips. Bronze-grey above. White below. Teeth: 14/14 in each half-jaw.

HABITAT: Open sea, throughout tropical belt, rarely coastal. Catches of this species (on long lines) increase the farther one gets from the coast. The most abundant of the oceanic sharks together with the Silky and the Blue. It is solitary (does not live in groups).

FOOD: Standard oceanic fish, barracuda, marlin, turtles, corpses of whales. Has been seen swimming at surface with mouth wide open among a sardine shoal being pursued by tunnies, only closing its mouth when a tunny swam headlong into it. A real nuisance to fishing boats, which it persistently follows, tearing the nets and devouring the catch (in particular whales hitched to the sides of ships). Attacks swimmers and boats. Always circles around divers. Was responsible for the slaughter among the survivors of the *Nova Scotia* when it sank off Natal. Very aggressive. The only chance for victims is the slowness of its approach.

MAXIMUM SIZE: 3.95 metres.

BLACKTAIL REEF SHARK
Carcharhinus wheeleri

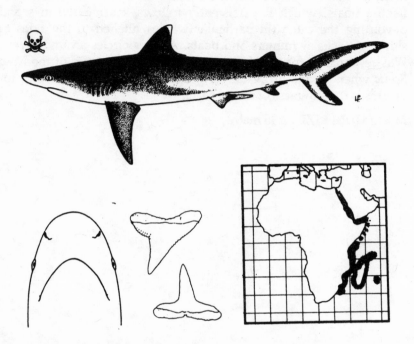

ORDER: Carcharhiniformes.

FEATURES: Grey and white, trailing edge of caudal fin black.

HABITAT: Common on the eastern coasts of Africa. No attack known, but aggressive during undersea fishing with harpoon or gun. Potentially dangerous.

MAXIMUM SIZE: 1.93 metres.

BLACK-TIP REEF SHARK
(Whaler, Black-tip Shark)
Carcharhinus melanopterus

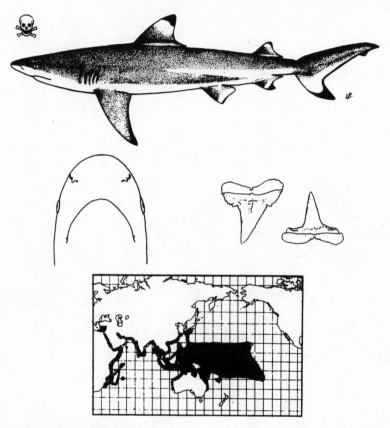

ORDER: Carcharhiniformes.

FEATURES: Brown-grey, tips of fins black, accentuated especially on the first dorsal by a whitish inner band. Short caudal fin.

HABITAT: See map. Rare or absent in whole eastern part of East Pacific (Marquesas, Pitcairn, etc.). Only in shallow waters, even as little as 30 centimetres. Found only in east of the Mediterranean, which it colonised via the Suez Canal. Encountered singly or in small groups; dangerous to divers carrying fish, but can be repelled fairly easily (less dangerous than the Grey Reef Shark). This is the shark encountered most by swimmers in the tropical Indo-Pacific. Many non-fatal attacks on legs apparently mistaken for normal prey. This is why the Marshall islanders swim between islands rather than walk. Diving into the water can be enough to alarm this fish.

MAXIMUM SIZE: No more than 2 metres (big groupers domineer the Black- tip Reef).

DUSKY SHARK (Whaler Shark)
Carcharhinus obscurus

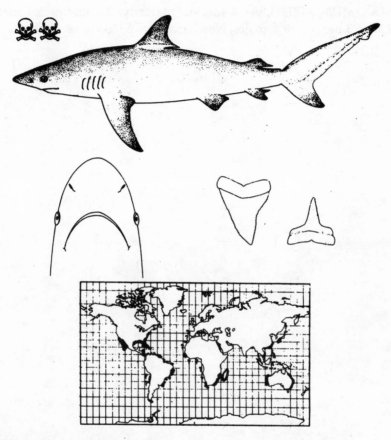

ORDER: Carcharhiniformes.

FEATURES: An all-grey shark with darker fins but no white nor black.

HABITAT: Coastal and pelagic. From the surf to depths of 400 metres. Often follows boats. Migrates north in summer and south in winter.

FOOD: Reef animals, from the bottom and the open sea, but only exceptionally refuse and dead bodies (unlike the Tiger and the Bull Shark). Sometimes confused with the Galapagos Shark, few attacks can be attributed to it but it is big in size. Its behaviour with regard to swimmers, surfers and divers is dangerous. This species has decreased along the Natal coast owing to the anti-shark nets, and adults have been replaced by young, which are sought-after prey of other large predators (Great White, Sand Tiger, Tiger).

MAXIMUM SIZE: Over 4 metres. Maturity: 2.8 metres for males, 2.55 to 3 metres for females. Newborn pups 69 to 100 centimetres.

TIGER SHARK

Galeocerdo cuvieri

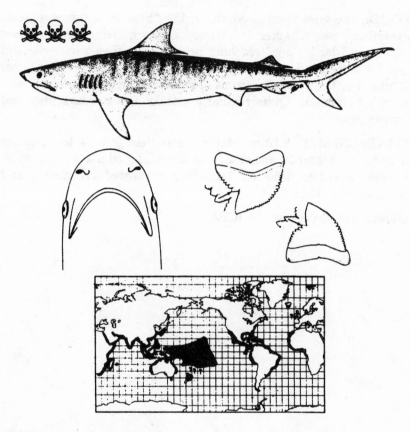

ORDER: Carcharhiniformes.

FEATURES: Caudal fin thin and pointed. Characteristic dark vertical stripes, especially on the young.

HABITAT: Coastal and pelagic, but operates over a wide range. All temperate and tropical seas. Highly tolerant of different habitats, but prefers murky waters and volcanic islands where heavy precipitation produces a high density of prey. Often in estuaries, ports, near piers or embankments, around atolls and lagoons. Not really oceanic, but capable of long crossings between islands. Apparently nocturnal. In the daytime, withdraws to deeper waters or hollows and waits for nightfall. The adult is solitary and resident or semi-resident around certain oceanic islands.

REPRODUCTION: It is the only ovoviviparous member of the Carcharhinidae family. Reaches maturity at 4 to 6 years, lives for 12 years minimum.

FOOD: The least specialised shark: the "hyaena of the seas", eats everything (see Chapter 4). Active, a strong swimmer, it cruises slowly and lazily, until reaching prey, and is then very agile and capable of tremendous acceleration. Together with the Bull and Great White, it is the most dangerous (particularly in the Caribbean and French Polynesia). Quite regularly attacks and devours man and attacks boats.

MAXIMUM SIZE: 9.10 m. Majority less than 5 m., a few females over 5.5 m. Maturity: males 2.26 to 2.9 m., females 2.5 to 3.5 m. A female caught in 1957 off Indochina measured 7.4 metres and weighed 3,110 kilos.

GENERAL: Recognised by IGFA.

GANGES SHARK

Glyphis gangeticus

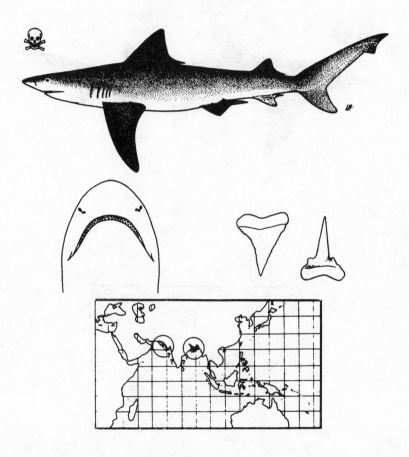

ORDER: Carcharhiniformes.

HABITAT: Riverine, fresh waters. Poorly known. Ascends Ganges, and has a reputation as a man-eater. Confused with Bull Shark, an established man-eater also in Ganges, and probably responsible for the numerous victims there. This shark's teeth are designed more for impaling fish than for dismembering mammals.

MAXIMUM SIZE: 2.04 metres.

SICKLEFIN LEMON SHARK

Negaprion acutidens

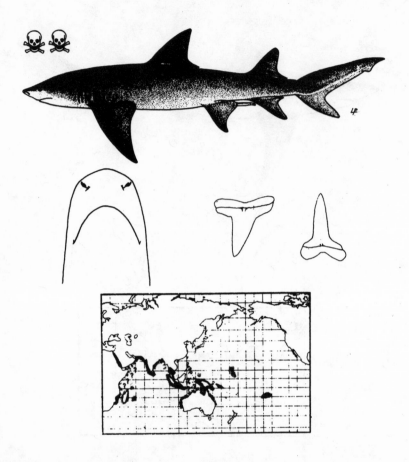

ORDER: Carcharhiniformes.

FEATURES: Pale yellow-brown. Said to be a sluggish species.

HABITAT: Coastal. On or near the bottom. Rarely at the surface except if it is stimulated by a prey. Prefers bays, estuaries, sandy bottoms, shallow waters to the extent that its fin breaks the surface.

FOOD: Bottom-living fish and sunfish. Timid, but will attack when knocked or accosted by a boat. One diver had to spend several hours on a rock, besieged by one of these sharks circling around him non-stop.

MAXIMUM SIZE: 3.10 metres.

LEMON SHARK
Negaprion brevirostris

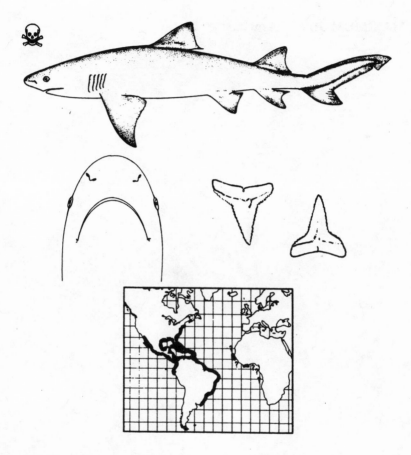

ORDER: Carcharhiniformes.

FEATURES: Pale brown-yellow.

HABITAT: Abundant Latin American coasts and Caribbean. Occasionally ventures out to sea, at or near surface, for the purposes of migration. Florida seems to be a preferred breeding area. Active night and day but more at dawn and dusk, like terrestrial predators. Seems to be territorial, but the adult moves around within an area of 300 square kilometres. It is adapted to environments with a low oxygen factor, exists in mangroves where temperature is high and micro-organisms plentiful. Adapts well to captivity. Has attacked man and boats, often after being harpooned or taken on a line. Its diet does not include any mammals and it usually does not attack if not attacked first, but its size, teeth and the power of its jaw demand that it be respected as dangerous. It is attracted to underwater fishing activities.

MAXIMUM SIZE: 3.4 metres.

BLUE SHARK
Prionace glauca

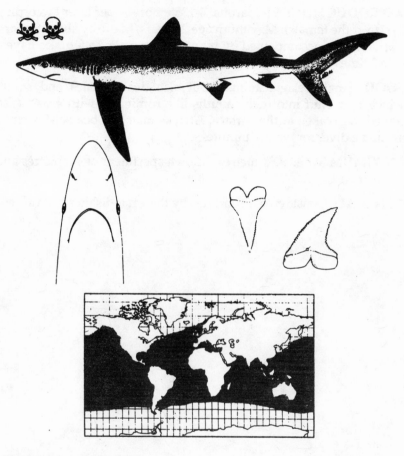

ORDER: Carcharhiniformes.

FEATURES: Blue above, bright blue on sides and changing abruptly to white on flanks. Pectoral fins long and narrow.

HABITAT: Entire world – the most cosmopolitan shark. Oceanic and in deep-water areas, ranging widely. Ventures on to coasts particularly at night. Often in large groups, close to surface, and in temperate waters. Prefers cold waters from 7– 16°C up to 21°C. So in the tropics, lives at depths of 80-220 metres, at 12-25°C. Can leap out of water when caught on a line. Often circles and watches prey before devouring it. Is capable of very rapid acceleration. Abundant in Pacific between 20°N and 50°N with varying numbers moving north in summer. Tagging demonstrated transatlantic migrations following the currents (Gulf Stream to Europe and North Equatorial Current to the Caribbean).

REPRODUCTION: Viviparous. 4-135 embryos per litter (according to size of the female). Maximum age 20 years. Maturity about 5 years. Copulation is accompanied by biting by the male, so females have a skin 3 times thicker than males.

FOOD: Small pelagic animals but also dead whales and squid, which they stuff into their mouths like ruminants with grass. Also tinned food taken at the surface. Attacks man and boats. May circle around a diver for twenty minutes.

MAXIMUM SIZE: 3.83 metres. Known specimens of 4.8 metres and 6.5 metres.

GENERAL: Considered a sport fish by the game-fishing authorities.

SMOOTH HAMMERHEAD

Sphyrna zygaena

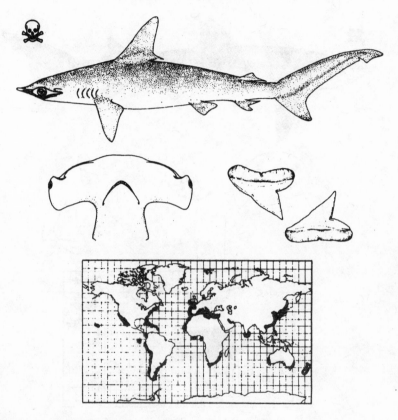

ORDER: Carcharhiniformes.

FEATURES: Dark olive, white below.

HABITAT: See map. Occurs in Europe. Coastal, pelagic and semi-oceanic. From surface level to 20 metres depth. May gather in enormous shoals. Content with temperate waters.

FOOD: Bony fish, cephalopods, skate., etc. Considered dangerous, but perhaps confused with the Great Hammerhead.

MAXIMUM SIZE: 3.7-4 metres.

WHITE-TIP REEF SHARK
Triaenodon obesus

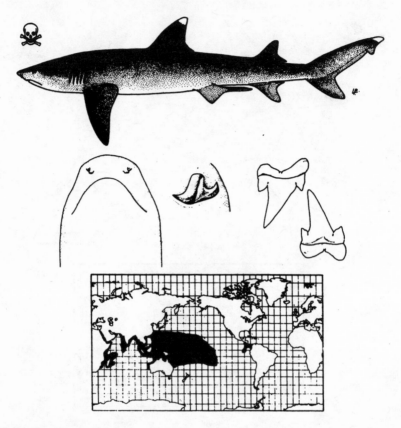

ORDER: Carcharhiniformes.

FEATURES: Grey with very visible white spots on first dorsal and caudal fin (see Chapter 4). Snout very short with prominent nostrils. Sardonic look to head face on.

HABITAT: See map. East Pacific: only Galapagos and Panama-Costa Rica. Very common in clear waters around coral reefs. Rests on bottom or in winding channels. In Oceania, one of the most widespread of the reef sharks together with the Grey and the Black-tip Reef Sharks. Active mainly at night, but comes out of daytime lethargy to inspect a boat that has just dropped anchor or a fish that has just been speared. Over one year movements of from 0.3 to 3 kilometres have been measured.

GENERAL: Maturity at 5, maximum age 25 years. Not suited for pelagic predation and specialises in benthic prey. It can itself be prey of large groupers and of bigger sharks. Rather timid: divers feed it by hand and have been attacked only when they had wounded it or were contending with it for a fish. More dangerous is the risk of ciguatera (a tropical disease) contracted by eating its flesh.

MAXIMUM SIZE: 2.13 metres.

GREAT HAMMERHEAD
Sphyrna mokarran

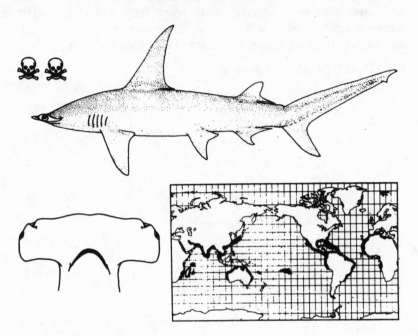

ORDER: Carcharhiniformes.

FEATURES: Easily recognised. Grey-brown above, pale below. No particular markings. 197 to 212 vertebrae.

HABITAT: Coastal and pelagic. Semi-oceanic. May enter water depths of a metre. Likes coral reefs. Nomadic and migratory.

REPRODUCTION: Viviparous. 13 to 42 embryos. Gestation period at least 7 months.

FOOD: Varied but especially rays, the spines of which often remain stuck in its mouth though without troubling it (one had 50 spines in its mouth, its throat and its tongue!).

MAXIMUM SIZE: 5.5 to 6.1 metres. Maturity around 2.5 metres.

GENERAL: Often confused with one of its congeners, it must be approached with respect and caution.

SAND TIGER SHARK
(Grey Nurse Shark, Spotted Ragged-tooth Shark)
Eugomphodus taurus

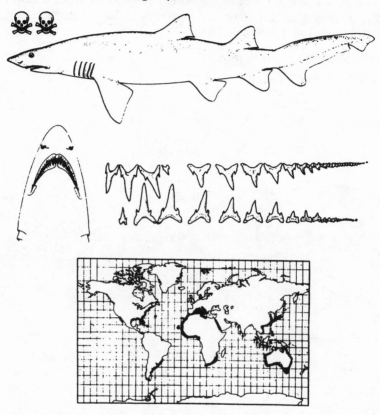

ORDER: Lamniformes (mackerel sharks).

FEATURES: No nictitating membrane, long mouth, big teeth with lateral points, pale brown with darker spots. Eye yellow.

HABITAT: Coastal shark. Strong swimmer, mainly nocturnal. Denser than water, it swallows air at the surface, retaining it in its stomach to give it neutral buoyancy. Lives solitarily or in groups. Seasonal migrant with the temperature.

REPRODUCTION: Ovoviviparous, with cannibalism *in utero*. 16 to 23 eggs but a single pup emerges after having eaten all the others inside the uterus; its size is then 1 metre.

FOOD: Very voracious. All bony fish, and small sharks, rays, crabs,

293

crayfish. Reputation as a man-eater in Australia, but is no doubt confused with other species. Not aggressive when not provoked, but it can attack underwater fishermen. Too many divers armed with powerheads delight in killing this easy prey as a hunter would kill a cow, causing a local decline in the species. A large number of attacks are attributed to it in South Africa.

MAXIMUM SIZE: 3.18 metres.

MEGAMOUTH SHARK
Megachasma pelagios

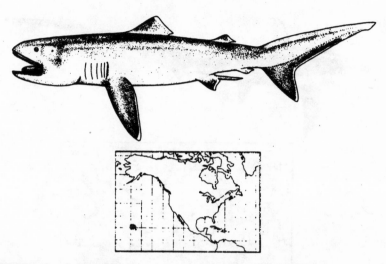

ORDER: Lamniformes.

FEATURES: Highly distinctive. Enormous mouth with numerous small teeth. One of the three shark species with a food filter (like Whale and Basking Sharks). First specimen discovered in 1976 in Hawaii, at a depth of 1.63 metres. Its stomach was full of shrimps 3 centimetres long when it got caught up in a pair of parachutes being used as sea anchors by a US Navy research vessel. Skin soft and with numerous wound marks from the Cookie-cutter Shark.

HABITAT: This is the only shark whose journeys between the surface and the depths and back are perfectly scheduled: ascent to the surface at sunset (18.00 hours) following the micro-organisms; drifting at the surface all night with mouth open to swallow shrimps and jellyfish; at dawn (06.00 hours) the shark descends again to about 150 metres, along with its prey.

SIZE: 4.46 metres. Another was caught in 1990, in California: 4.5 metres and 900 kilos. So far 5 specimens have been captured (Australia-Japan), all males of about 1 tonne, with the whereabouts of the females remaining unknown.

THRESHER SHARK (Fox Shark)
Alopias vulpinus

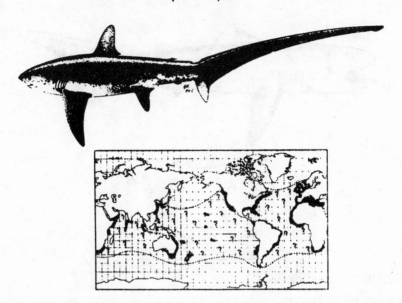

ORDER: Lamniformes.

FEATURES: Caudal lobe as long as the body, abdomen white. 29 rows of small teeth. Note the very marked development of apophyses (processes) of the vertebrae at the tip of the tail, making it a veritable bludgeon for stunning its prey.

HABITAT: Coastal in deep-water areas. Active, and strong swimmer.

REPRODUCTION: Ovoviviparous; cannibalism in womb.

FOOD: Small fish in shoals: mackerel, garfish, etc. Uses its tail to gather together and stun its prey. Often caught by its tail on long lines. No danger to swimmers, it is said to have attacked boats. With one lash of its tail it decapitated a fisherman on the deck of an American trawler in the North Atlantic. Has never attacked divers even though it circles around them just within sight.

MAXIMUM SIZE: 5.49 metres. Newborn pups from 1.14 to 1.5 metres.

GENERAL: Approved by IGFA.

BASKING SHARK
Cetorhinus maximus

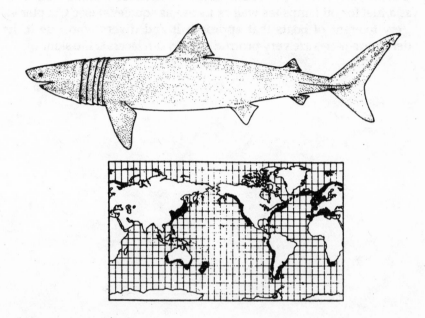

ORDER: Lamniformes.

FEATURES: Enormous gills which almost completely encircle the head.

HABITAT: Pelagic coastal, in boreal as well as warm temperatures. Often at the surface alone or in a shoal of about a hundred. Two or three may follow each other and may in the past have been taken for a single gigantic sea serpent. Sometimes leaps out of the water, possibly to rid itself of its many parasites, which include lampreys. These parasites cannot, however, penetrate the tremendous armour of denticles covering this shark. Considerable seasonal migrations.

REPRODUCTION: Ovoviviparous, with cannibalism in the womb. Gestation period 3 years. Maturity around 6 or 7 years.

FOOD: Exclusively plankton, of which its stomach can hold 500 kilos. An average adult can filter 2,000 tonnes of water per hour at a rate of 2 knots, by keeping its mouth and gills wide open. Its liver allows it possibly to hibernate in winter for the period when it loses its filters and when plankton is very scarce.

MAXIMUM SIZE: 12.2 to 15.2 metres. Maturation at 8.50 metres for females.

GENERAL: The oil from its liver can be used for tanning leather, or as a fuel for oil lamps (as well as its use as squalene) (see Chapter 4). Very tolerant of boats that approach it and divers who ride it. Its dermic denticles are very protruding and can lacerate the skin.

GREAT WHITE SHARK
(White Pointer, Blue Pointer)
Carcharodon carcharias

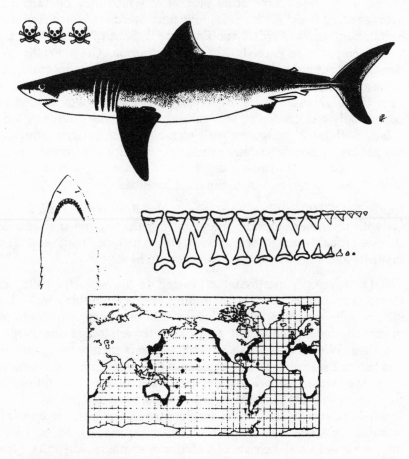

ORDER: Lamniformes.

FEATURES: Fairly long conical snout; large triangular blade-like teeth, serrated and as sharp as a razor. Long gills, big dorsal fin, ventral surface white.

HABITAT: Coastal and temperate, but occasionally makes deep-water incursions out to the ocean. Often in the surf zone and even in very shallow waters, but prefers islands, in particular those which are inhabited by colonies of pinnipeds. Has been caught on a long line at depths down to 1,280 metres, but often at the surface as well. A very agile, powerful shark, swimming like a tunny, which enables it to cruise for a very long time at relatively slow speed; has been recorded, by telemetry, at a mean speed of 3.2 knots over 190 kilometres in 2 days. Capable of sudden accelerations and sometimes leaps out of the water. It seems that only the biggest individuals have a very wide temperature bracket in which they operate. The average-sized ones (2.8 metres) move northwards in North America, in summer, and towards Cape Province in South Africa, when the water temperature exceeds 22°C. In California, Great Whites are present throughout the year (in increasing numbers according to some studies), slightly more so when the water changes from 11°C to 14 or 15°C. This shark is often solitary or in twos, although occasionally a dozen or so will gather near a major source of food (island with birds, penguins, seals etc), but never in large groups. It has not been possible to demonstrate territoriality, but certain events have proved that the same individual was returning, sometimes for several years in succession, to the same localities.

REPRODUCTION: Ovoviviparous, with intra-uterine cannibalism (as with the other lamnids). Pregnant females are never caught as there is probably female segregation, and, when they reach sufficient maturity and size, they are too enormous to be captured.

FOOD: A real "superpredator" owing to its size, its teeth, its metabolism and its efficient swimming. It eats everything including other sharks, dolphins, turtles (less so than the Tiger Shark), and all mammals, dead or alive, but it swallows fewer strange objects than the Tiger. When bigger than 3 metres, it feeds mainly on mammals and fish of 2 metres or less. Considered the most dangerous shark in temperate waters where it does not have many competitors. In tropical waters, it is likewise regarded as the most dangerous, but the Tiger Shark and the Bull Shark are close behind it, even surpassing it in some areas (the Bull Shark in Natal, for example). Most attacks have occurred on the coasts of California, southern Australia, New Zealand and South Africa. The majority of these fortunately involve just a single bite, explaining a not insignificant victim survival rate. This type of aborted attack has been attributed to non-feeding aggression, or to a mistake in identification, or else to a tasting test of a potential prey. A sort of "play" has also been mentioned.

MAXIMUM SIZE: From 6.40 to 8 metres.

GENERAL: Recognised by IGFA.

SHORTFIN MAKO (Blue Pointer)
Isurus oxyrinchus

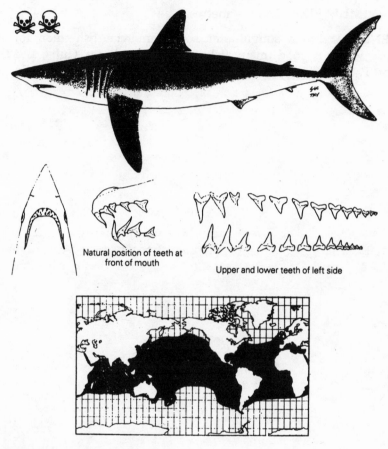

Natural position of teeth at front of mouth

Upper and lower teeth of left side

ORDER: Lamniformes.

FEATURES: Long, narrow pectoral fins, minute second dorsal fin, pointed conical snout. White below.

HABITAT: Coastal and oceanic, temperate and tropical. Common and very active. Rarely in waters below 16°C. Possibly the fastest of the sharks and one of the most agile. Famous for its leaps, making jumps of several times its own length out of the water. Capable of exceptional top speeds if caught on a hook or when in hunting phase.

FOOD: Eats all other fish and other sharks, but only rarely mammals (porpoises). Dangerous, but its habitat far offshore means that contact with swimmers is rare. Should *never* be harpooned or provoked underwater, for it then becomes a veritable torpedo. Much too swift for an anti-shark weapon to be put to use. Second after the Great White in attacking boats, especially those of game fishermen.

MAXIMUM SIZE: 3.94 to 4 metres.

GENERAL: Much sought-after by commercial fishermen for its flesh, which is good (many fisheries in Italy, Africa, Cuba, and the west Pacific). Highly prized in game-fishing. Recognised by IGFA.

GOBLIN SHARK

Mitsukurina owstoni

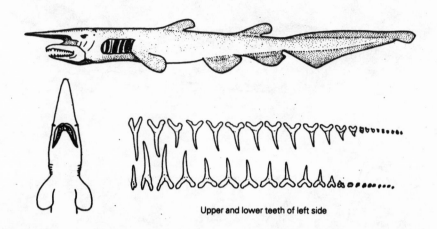

Upper and lower teeth of left side

ORDER: Lamniformes.

FEATURES: Highly characteristic.

HABITAT: Bottom-dwelling shark (at least 5.5 metres depth), forehead "ram" probably for throwing up its prey, specialised teeth for rapid projection of the head and piercing of the prey. Very rarely known. Rare (Bay of Biscay, Portugal), South Africa, southern Australia, Japan and the Guianas.

MAXIMUM SIZE: 3.35 metres.

PORBEAGLE

Lamna nasus

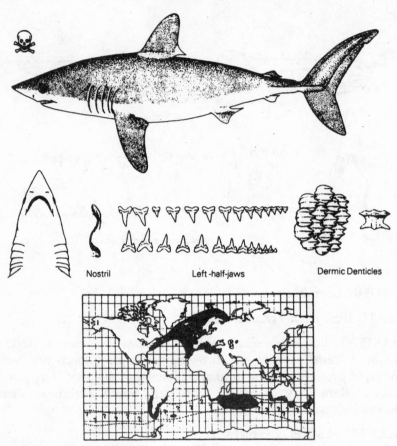

Nostril   Left-half-jaws   Dermic Denticles

ORDER: Lamniformes.

HABITAT: Littoral and in deep water. Prefers cold waters below 18°C. Absent from equatorial waters. Agile in pursuing prey, but puts up little resistance when taken on the hook (the opposite of the Mako). Near coasts and at the surface only in summer.

REPRODUCTION: Ovoviviparous and with cannibalism in the womb like all the lamnids. Maturity at five years. Lifespan 20 to 30 years.

FOOD: Small fish. A few attacks on man and on boats.

MAXIMUM SIZE: 3 to 3.7 metres.

GENERAL: Heavily fished, especially in France for its flesh, its fins and its oil. Widely used in England for the traditional fish and chips. Sold in French supermarkets under the name of "veau de mer" (sea veal). Approved by IGFA.

ZEBRA SHARK
Stegostoma fasciatum

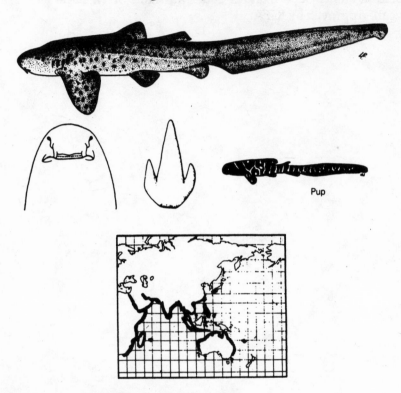

Pup

ORDER: Orectolobiformes (carpet sharks).

FEATURES: Tail as long as body. Colour very variable with age.

HABITAT: Very widespread on coral reefs. Sluggish. More active at night. Rests on the sandy bottom near coral.

REPRODUCTION: Oviparous.

FOOD: Molluscs, crustaceans and small bony fish.

MAXIMUM SIZE: 3.54 metres.

GENERAL: Not dangerous.

NURSE SHARK

Ginglymostoma cirratum

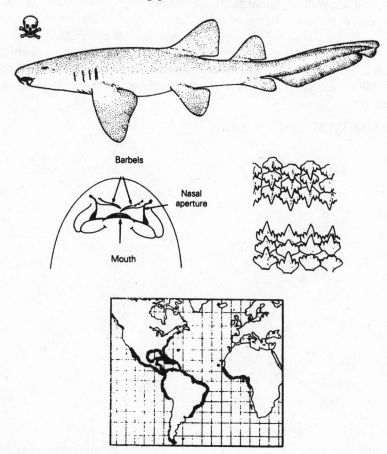

ORDER: Orectolobiformes.

FEATURES: Fairly long barbels. Yellow-brown or grey-brown.

HABITAT: Coastal sea beds. Sluggish. Remains on bottom throughout the day, in shallow water. Sometimes stacked one on top of the other in groups of 3 to 30 individuals. Active at night. Can crawl on the bottom using its muscular pectoral fins as "legs".

REPRODUCTION: Ovoviviparous: 21 to 28 eggs.

FOOD: Eats bottom-dwelling invertebrates and occasionally seaweed. Its mouth allows it to take in shellfish by suction. Not aggressive, but several unprovoked attacks on divers and swimmers. Some have bitten and refused to let go, to the extent that instruments were necessary to free the victim. In an attack on one diver, the shark bit him in the chest and then wrapped its two pectoral fins around him (the sex was not identified). Of the many people who try to ride it, spear it or walk on top of it, some become victims of their own foolhardiness. This shark's teeth are small, but the muscles of the jaws make them real vices. Record in captivity: 25 years; this enabled them to be studied and many experiments to be made, for they are capable of learning. Its skin is much sought-after, being as tough as armour.

MAXIMUM SIZE: 4.3 metres.

SPOTTED WOBBEGONG
Orectolobus maculatus

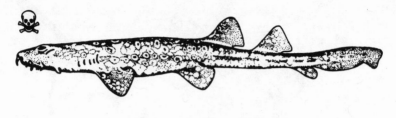

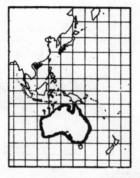

ORDER: Orectolobiformes.

FEATURES: 8 to 10 lobes of skin on each side of the head.

HABITAT: Abundant bottom-living shark of temperate to tropical seas. Sluggish, camouflaged against the sea bed. Apparently nocturnal. Often comes back to the same locality.

REPRODUCTION: Ovoviviparous: up to 37 embryos.

MAXIMUM SIZE: 3.2 metres.

GENERAL: It bites if somebody steps on its back or near its mouth, and a fisherman lost a foot this way. Unprovoked attacks have been reported, on people and also boats around Australia. Should be treated with respect. Its cousin, the Bearded Carpet Shark (same size), is said to have attacked and killed the Papuans of New Guinea.

WHALE SHARK

Rhiniodon typus

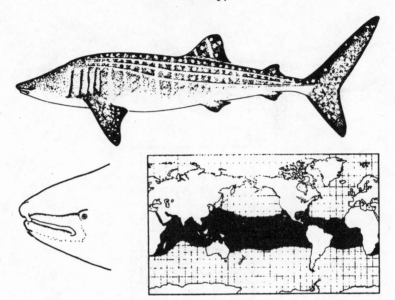

ORDER: Orectolobiformes.

FEATURES: Huge flattened head with mouth at end, very large number of teeth and unique filtering system on the inside of its gills. Several protruding ridges on its sides (see Chapter 4).

HABITAT: Deep-water areas and coastal. Throughout tropical and temperate belt. May enter lagoons. Often at the surface singly or in groups of several hundreds. Prefers surface areas of 21 to 25°C and salinities of 34 to 34.5% (optimum for planktonic and nektonic development).

REPRODUCTION: Oviparous or ovoviviparous, it is not known which (an egg found in 1953 was 30 centimetres long but, on the other hand, the smallest Whale Shark encountered had an umbilical scar).

FOOD: Plankton and necton, plus a few small fish (tunny, sardine, anchovy, mackerel etc). Feeds at the surface, often in a vertical position with its mouth out of the water. This shark does not depend on forward movement to feed (unlike the Basking Shark). Not aggressive, but may react against being hit by a boat.

MAXIMUM SIZE: 18 metres.

PORT JACKSON SHARK
Heterodontus portus jacksoni

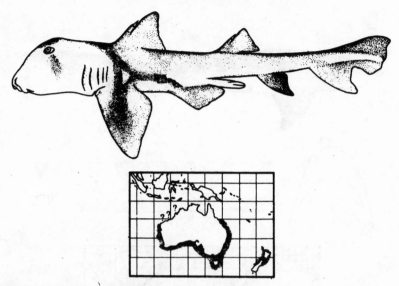

ORDER: Heterodontiformes.

HABITAT: Coastal bottom-dwelling shark, nocturnal.

REPRODUCTION: Oviparous. Eggs 13 to 17 centimetres long and 5 to 7 centimetres across, with thread-like appendages for securing an anchor hold. Hatch after 9 to 12 months.

FOOD: Seasonal. Feeds mainly on benthic invertebrates. Not a danger but, if attacked, it may pursue a diver and bite him.

MAXIMUM SIZE: 1.65 metres.

GENERAL: The anatomy of this shark has not altered over 300 million years while its blood globin has observed the same changes as man's subjected to the natural variations of the environment. Kimura deduces from this that "nature" cannot have exerted any natural selection on the globin and therefore calls into question Darwin's theory of evolution, based on natural mutation-selections.

BROADNOSE SEVEN-GILL SHARK
Notorynchus cepedianus

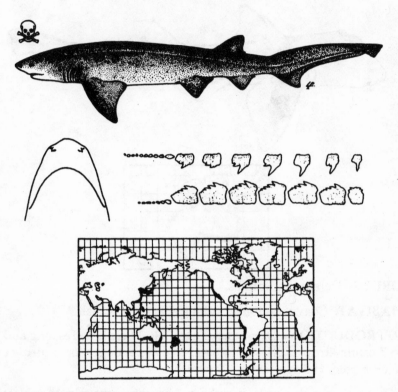

ORDER: Hexanchiformes.

HABITAT: Benthic, occasionally depths of 1 metre. Only on coasts.

FOOD: Eats anything, fish, sharks, dead bodies. Aggressive if provoked, considered potentially dangerous in open waters. Has attacked divers in captivity. Puts up strong and dangerous resistance and often has to be put out of its misery with a firearm before being hauled on board. Sought after for its flesh. Fished for from shore in South Africa.

MAXIMUM SIZE: 2.90 metres.

GLOSSARY

Benthic: Relating to the benthos. The benthos is the community of aquatic organisms which live in sea beds and depend on them for their subsistence (the benthos is the opposite of plankton).

Elasmobranchs: Group of fish whose skeleton is cartilaginous, as opposed to the group of teleosts or bony fish. The elasmobranchs include the sharks, the rays and skates, and the sawfish.

Endorphins: Hormonal substances secreted *in situ* in certain areas of the brain, in particular in the pain centre. A substance having the properties of morphine is secreted in this way at times of acute stress.

Endurance: An endurance effort is an effort of moderate but prolonged intensity. It requires oxygen and gets the red muscle fibres working. The shark is handicapped on the endurance level owing to its small heart, variably developed red muscles, and a variable haematocrit.

Homeotherm: Refers to animals whose temperature is constant and not influenced by the environment (owing to thermoregulation).

Midships frame, amidships: The widest cross-section of a ship's hull or an aircraft's fuselage; by extension, in this book, the broadest section of the body of a whale.

Nekton: Small aquatic organisms that are free-swimming in the waters of the sea (compare with plankton).

Nycthemera: Period of time covering one day and one night and corresponding to a biological cycle.

Pelagic: Relating to the high sea. Animal that lives in the deepest parts of the sea.

Pinnipeds: Of the order of mammals adapted to an aquatic life and covered with fur: walrus, sea lion, seal, etc. Favoured prey of sharks (Great White).

315

Plankton: Community of organisms (of very small size) which live suspended in the water of the sea.

Poecilotherms: Cold-blooded animals whose temperature is variable: reptiles, fish (sharks), etc.

Predator: Animal which feeds on other animals which it kills in order to survive.

Resistance: A resistance effort is an intense but brief effort, capable of being made without oxygen for a few tens of seconds, and setting the white muscle fibres in operation. The shark is capable of very great resistance efforts when accelerating or making attacking charges (the opposite of "endurance" effort).

BIBLIOGRAPHY

RICHARD BACKUS AND LINEAWEAVER, *The Natural History of Sharks*, André Deutsch.

ASEY AND PRATT, *Biology of the White Shark*, Southern California Academy of Sciences, 1985.

LEONARD COMPAGNO, *Sharks of the World*, vol. 4, 655 pp., FAO Species Catalogue, Rome, 1984.

LEONARD COMPAGNO, *Shark Attack*, 1989 (unpublished study).

VICTOR COPPLESON AND PETER GOADBY, *Shark Attack*, Angus Robertson Publishers, Australia, 1988.

GULF OF MEXICO FISHERY MANAGEMENT COUNCIL, *Fishery Management Plan for Sharks and Other Elasmobranches in the Gulf of Mexico*, 1979.

DAVID H. DAVIES AND PIETER MARITZBURG, *About Sharks and Shark Attack*, Shuter and Shooter.

PERRY W. GILBERT, *La vie du requin*, 1947.

GRUBER, SLOTKIN AND NELSON, *Shark Repellents: behavioural bioessays in laboratory and field*, Springer Verlag, Berlin, 1984.

OCEANOGRAPHIC RESEARCH INSTITUTE, *Reactions of the Sharks to Gill Net Barriers under Experimental Conditions*, Durban, 1972.

MACCORMICK, ALLEN AND CAPTAIN YOUNG, *Shadows in the Sea*, Sidgwick and Jackson, London.

MYRBERG, *Understanding Shark Behavior*, Miami.

NELSON, JOHNSON AND PITTENGER, *Agonistic Attacks on Divers and Submersibles by Sharks: antipredatory or competitive?*, Bulletin of Marine Science, 1986.

R. PATERSON, *Shark Prevention Measures Working Well*, Australian

Fisheries, Australia, 1986.

PATRICIA POPE, *A Dictionary of Sharks*, Great Outdoors Publishing Co., Florida, 1977.

PORTLAND, *Sharks*, Oregon State University, 1985.

JOHN RANDALL, *Sharks of Arabia*, Immel Publishing, London, 1986.

NEW YORK UNIVERSITY, *The Ichthyotoxic Mechanism of Pardaxin*, 1983.

NORMAN WYNNE, *Facts on Sharks*, Natal Shark Board, 1988.